# MOLECULAR PHOTOBIOLOGY
*Inactivation and Recovery*

# Molecular Biology

## An International Series of Monographs and Textbooks

### Edited by

**BERNARD HORECKER**
Department of Molecular Biology
Albert Einstein College of Medicine
Yeshiva University
Bronx, New York

**NATHAN O. KAPLAN**
Department of Chemistry
University of California, San Diego
La Jolla, California

**JULIUS MARMUR**
Department of Biochemistry
Albert Einstein College of Medicine
Yeshiva University
Bronx, New York

**HAROLD A. SCHERAGA**
Department of Chemistry
Cornell University
Ithica, New York

HAROLD A. SCHERAGA. Protein Structure. 1961
STUART A. RICE AND MITSURU NAGASAWA. Polyelectrolyte Solutions: A Theoretical Introduction. *With a contribution by Herbert Morawetz.* 1961
SIDNEY UDENFRIEND. Fluorescence Assay in Biology and Medicine. Volume I – 1962. Volume II in preparation – 1969
J. HERBERT TAYLOR (Editor). Molecular Genetics. Part I – 1963. Part II – 1967
ARTHUR VEIS. The Macromolecular Chemistry of Gelatin. 1964
M. JOLY. A Physico-chemical Approach to the Denaturation of Proteins. 1965
SYDNEY J. LEACH (Editor). Physical Principles and Techniques of Protein Chemistry. Part A – 1969. Part B in preparation
KENDRIC C. SMITH AND PHILIP C. HANAWALT. Molecular Photobiology: Inactivation and Recovery
*In preparation*
RONALD BENTLEY. Molecular Asymmetry in Biology. Volumes I and II
JACINTO STEINHARDT AND JACQUELINE A. REYNOLDS. Multiple Equilibria in Proteins

# MOLECULAR PHOTOBIOLOGY
Inactivation and Recovery

## Kendric C. Smith
Department of Radiology
Stanford University
School of Medicine
Stanford, California

## Philip C. Hanawalt
Department of Biological Sciences
Stanford University
Stanford, California

ACADEMIC PRESS   New York and London   1969

ACADEMIC PRESS, INC.
111 Fifth Avenue, New York, New York 10003

United Kingdom Edition published by
ACADEMIC PRESS, INC. (LONDON) LTD.
Berkeley Square House, London W.1

LIBRARY OF CONGRESS CATALOG CARD NUMBER: 69-18345

PRINTED IN THE UNITED STATES OF AMERICA

# Preface

The sun is the principal source of energy for the biosphere. It figured predominantly in the steps of organic evolution that culminated in life on earth, and we now depend upon its radiations for our continued existence. The beneficial effects of sunlight are highly specific; they are mediated through very specialized organic molecules that have evolved for the evident purpose of trapping the sun's energy. Yet, it is painfully apparent to the sunbather that not all of the effects of light are beneficial. Most of the photochemical reactions that occur in biological systems are deleterious to life.

In this book we consider those interactions of light, particularly in the ultraviolet region of the spectrum, that result in damage to proteins and nucleic acids in living systems. We discuss the kinds of photochemical reactions that can occur and the effects such photochemistry may have at the molecular, cellular, and organismal levels. Within recent years it has been shown that cells can repair their damaged genetic material, and thus recover from the otherwise inactivating effects of light. These recovery mechanisms, which have been elucidated largely through research in photobiology, are considered in detail in this work. Finally, some biological effects of ionizing radiation are compared and contrasted with those of ultraviolet radiation.

This book is the outgrowth of a course that we have given at Stanford University since 1963. The students have been largely upper division science majors and graduate students in disciplines ranging from physics to biology. With the rapid advances in the chemistry of photon effects on molecules and the equally rapid elucidation of the intricacies of biological mechanisms it has become difficult for a common language to unite the several areas of

v

inquiry. We have attempted to bridge this communication gap in photobiology between physicist, chemist, and biologist. Where technical terms in these disciplines are required for clarity they have been defined in a glossary.

Ultraviolet photobiology can be used as a powerful and sensitive probe for studying the intracellular machinery. The photon probe has the advantage that it can enter the cell to pinpoint a target molecule without the complications of permeability problems and the side reactions that occur when various chemical probes are used. We have tried to emphasize some of the important ways in which photon probes can be used to study biochemical mechanisms; we hope that this book will stimulate increased research interest in this field.

This work was not designed as a comprehensive source of detailed reference material. Rather, it is an attempt to integrate the ideas and current understanding in the field for instructional purposes. Therefore, we have kept references to a minimum in order that the continuity of the text not be interrupted. General references for further study are included at the end of each chapter.

We are indebted to the students in our course who often encouraged us by listening attentively. Some of them also read and commented upon early drafts of these chapters. It is impossible to give appropriate credit to all of our colleagues who contributed to this work in helpful discussions. Finally, we are indebted to our wives, Marion and Jo, for their encouragement and patience during the final phases of completing the book.

KENDRIC C. SMITH
PHILIP C. HANAWALT

*March, 1969*

# Contents

# 7. RECOVERY FROM PHOTOCHEMICAL DAMAGE

# 8. ULTRAVIOLET MUTAGENESIS

# 9. PHOTODYNAMIC ACTION

## 10. COMPARISON OF UV AND IONIZING RADIATION

# Glossary

α-HELIX: the principal regular configuration that can be assumed by a protein, in which stability is maintained by intrachain hydrogen bonds between the peptide linkages that connect the amino acids in the chain.

ALLELE: one of two or more alternate forms of a gene. Thus, one could have an allele specifying blue eyes or one specifying brown eyes of the gene for eye color.

AMINO ACID ACTIVATING ENZYME: an enzyme which recognizes a specific amino acid and links it to the appropriate transfer RNA species, using energy from ATP to form an activated complex.

ANTICODON: the triplet of nucleotides in a transfer RNA molecule that serves to recognize the triplet of nucleotides that constitutes the codon in messenger RNA.

ANTIGEN: any substance which, when introduced into a vertebrate elicits the production of molecules (antibodies) which bind specifically to the inducing substance.

AUXOTROPH: a mutant organism that lacks a particular biosynthetic capability present in the parent (prototroph). (e.g., an arginine requiring auxotroph of *E. coli*.)

BACTERIOPHAGE: a virus that specifically infects bacterial cells.

β-FORM: a less common (than the α-helix) configuration for proteins in which the chains lie side by side in the so-called pleated sheet arrangement stabilized by interchain hydrogen bonds.

BROWNIAN MOTION: the motion of a particle in a fluid as a result of its bombardment by molecules in random thermal motion.

CARTESIAN COORDINATES: a system for locating a point on a plane by its distance from each of two intersecting lines, or in space by its distance from each of three planes intersecting at a point.

CHLOROPLAST: the chlorophyll-containing organelle that facilitates the conversion of photon energy into chemical energy in the process of photosynthesis.

xi

CHROMOPHORE:   the chemical group in a molecule that is capable of photon absorption.

CHROMATOGRAPHY:   the analytical (or preparative) procedure by which chemical compounds (e.g., amino acids) are separated on the basis of their differing solubilities in solvents in a supporting matrix.

CILIA:   fibers of locomotion on some unicellular protozoans, or functional organelles on certain differentiated cells in higher organisms.

CLEAVAGE:   the series of mitotic divisions in the early stages of the growth of an embryo from an egg.

CODON:   the triplet of nucleotides in a DNA or messenger RNA that codes for a particular amino acid (or signals the end of the message).

COMPTON EFFECT:   the interaction of a high energy photon with a single electron in an atom resulting in the ejection of the electron and the secondary emission of a lower energy (longer wavelength) photon.

CONJUGATED DOUBLE BONDS:   a bonding situation somewhat intermediate between that of the double bond and that of a single bond such that some properties of double bonds (e.g., restriction of rotation) are maintained.

COVALENT BOND:   a bond in which electrons are shared in a molecular orbital involving both the bonded atoms.

CUVETTE:   a sample receptacle for optical studies. A typical cuvette for UV spectrophotometry might be a rectangular quartz bottle with a light path of exactly 1 cm.

DENATURATION:   the change in configuration of a protein from its functional state to a less ordered (generally nonfunctional) state; the breaking of the hydrogen bonds connecting the two strands in a double-stranded nucleic acid.

DEOXYRIBONUCLEASE:   an enzyme that degrades DNA; either an endonuclease that produces breaks in the middle of a strand (or both strands) or an exonuclease that removes nucleotides from the ends of strands.

DIPLOID:   having each chromosome (except the sex chromosomes) represented twice.

DIPOLE:   a separation of equal positive and negative charges.

DNA POLYMERASE:   an enzyme that synthesizes DNA in the presence of a DNA template and appropriate precursors (i.e., deoxyribonucleoside triphosphates).

ELECTRON SPIN RESONANCE:   the characterization of material on the basis of the magnetic properties of its electrons.

ENERGY SINK:   a molecule or molecular group that readily accepts energy transferred from other components in the system.

EUCARYOTIC:   a cell type in which the nucleus is separated from the rest of the cell (the cytoplasm) by a discernable membrane.

EXONUCLEASE III:   a deoxyribonuclease that specifically removes nucleotides or phosphoryl groups sequentially from the 3′ end of one strand of a double-stranded nucleic acid.

FIBROBLAST:   a type of cell found in connective tissue in mammalian systems; characterized by irregular branching morphology.

GENE:   the unit of inheritance.

GENOME:   the total genetic complement of an organism.

GENOTYPIC:   referring to the hereditary potential of an organism (cf. phenotypic).

GERMINATION:   the beginning of growth or development

HAPLOID:   having each chromosome represented only once.

HEMAGGLUTINATION:   the aggregation of red blood cells.

HETEROCYCLIC:   refers to ring structures of the benzene type but which (like pyrimidines) contain other elements in addition to carbon in the ring.

HYDROGEN BOND:   a directed ionic bond involving a shared proton.

HYDROPHILIC:   charged or polar; interacts with a polar solvent such as water.

HYDROPHOBIC:   noncharged, nonpolar; interacts with other noncharged units more readily than with charged ones.

INTERPHASE:   the stage in the mitotic cycle during which the chromosomes are not evident, presumably because they have a more extended configuration than the condensed state which characterizes the other mitotic stages.

KINETOCHORE:   the point at which a chromosome is attached to the spindle apparatus for purposes of directing its motion during the mitotic process.

LOCUS:   position of a gene on a chromosome.

LOG PHASE OF GROWTH:   the phase of growth characterized by an exponential increase in the number of cells with time.

MARKER:   a gene of known function and known location on the chromosome.

MESSENGER RNA:   the carrier of the genetic message from DNA for protein synthesis.

MITOCHONDRIA:   organelles responsible for the conversion of chemical energy in pyruvic acid to ATP via the citric acid cycle.

MITOTIC CYCLE:   the sequence of steps by which the genetic material is equally partitioned prior to division of a cell into daughter cells.

MONONUCLEOTIDE:   a single nucleotide containing a purine or pyrimidine base, a sugar (ribose or deoxyribose), and a phosphate group.

MUTATION:   an inheritable variation in the genome.

NONCONSERVATIVE REPLICATION:   the mode of replication in which there is no net increase in amount of DNA but only a replacement of parental DNA with newly synthesized DNA.

NUCLEOSIDE TRIPHOSPHATE:   a precursor for nucleic acid synthesis and also the principal mode for transporting chemical energy about the cell, containing a purine or pyrimidine base, a sugar (ribose or deoxyribose), and three phosphate groups.

OLIGONUCLEOTIDE:   a chain of several nucleotides.

OPTICAL ROTATORY DISPERSION (ORD):   the measured variation in optical rotation as a function of wavelength. Useful for studying conformational changes in proteins and nucleic acids.

ORBITAL:   the probability distribution or wave function for an electron in a particular energy state in an atom or molecule.

OSMOTIC SHOCK:   a sudden change in pressure within a closed semipermeable membrane (such as the cell membrane). Osmotic shock can be sufficient to shear a large macromolecule such as DNA into smaller fragments.

PAIR ANNIHILATION:   the production of a photon upon the dissipation of energy when an electron and a positron collide.

PAIR PRODUCTION:   the synthesis of an electron-positron pair from photon energy.

PARAMAGNETIC:   a substance that enhances the magnetic flux density when placed in a magnetic field. Oxygen is paramagnetic, nitrogen is not.

PARTHENOGENESIS:   development of an unfertilized egg into an organism.

PETITE MUTANT:   a respiratory (i.e., mitochondrial) deficient mutant of yeast. These mutants typically form small colonies.

PHENOTYPIC:   referring to the observable properties of an organism produced by the interaction of its genotype with the environment.

PHOTOELECTRIC EFFECT:   the emission of electrons from a substance in response to the impingement of photons of sufficient energy.

PHOTOSYNTHESIS:   the process that couples photon energy to biosynthesis.

PHOTOTAXIS:   movement response of organisms to light.

PHOTOTROPISM:   directional growth response of organisms to light.

PLAQUE:   the "hole" in a layer of bacterial cells on an agar plate produced by the presence of bacteriophage infection in that region. Plaque counting can be used as an assay for viable phage in much the same manner that colony counts can be used to assay viable bacterial cells.

POISSON DISTRIBUTION: a probability distribution in which the variance equals the mean. If the total number of individuals observed under certain circumstances varies according to a Poisson distribution with mean $m$, the chance of obtaining $x$ individuals is

$$\frac{e^{-m}\, m^x}{x!}$$

where $x$ can be any whole number from 0 to $\infty$.

POLARIZABILITY: the tendency of a substance to have dipoles induced when it is placed in an electric field.

POLAR SOLVENT: solvent molecules containing charged groups or dipoles that can interact with other solvent molecules or with dipoles in the solute molecules.

POLYPLOIDY: having each chromosome (except the sex chromosomes) represented more than twice.

PROPHAGE: the state of a lysogenic bacteriophage in which its genome is integrated into the chromosome of the host.

PROPHASE: the early stage in the mitotic cycle in which the chromosomes condense and begin to move toward the equatorial plane. DNA synthesis and duplication of the chromosomal material has been completed by this stage.

PROPLASTID: the subcellular organelle that is the precursor for the chloroplast, as in *Euglena gracilis*.

PROTOTROPH: the parental or wild type organism from which nutritionally deficient *auxotrophs* may be produced by mutation.

RADICAL: a highly reactive chemical species having an unpaired electron.

RECOMBINATION: the appearance together in offspring of traits found separately in each of their parents.

REPLICATIVE FORM (RF): the double-stranded intermediate synthesized from the single-stranded genome of a virus or bacteriophage.

RNA POLYMERASE: the enzyme that copies the DNA nucleotide sequence in the synthesis of complementary RNA strands.

SATURATED BOND: the situation in which no further atoms (e.g., hydrogen) can be combined with the bonded atoms (e.g., carbons).

SEMICONSERVATIVE REPLICATION: the normal mode of DNA replication in which each of the parental DNA strands is conserved in a different daughter DNA molecule.

SEROLOGICAL GROUPING:   grouping of organisms according to common or cross-reacting antigens.

SINGLET STATE:   all electrons are present in pairs in which the spins are opposite so that the resultant spin for each pair is zero.

SPORE:   the dormant state which certain cell types enter when presented with environmental circumstances adverse to growth.

STATIONARY PHASE OF GROWTH:   the leveling off of growth in a culture of cells as the culture density approaches a limiting value or as certain nutrients are exhausted.

STREPTOMYCIN MARKER:   a gene that affects sensitivity of a cell to the antibiotic streptomycin.

SUPPRESSOR MUTATION:   a mutation that totally or partially restores a function lost by a primary mutation, but which is located at a genetic site different from the primary mutation.

TRANSFORMING DNA:   DNA purified from a strain of bacteria that carries some genetic trait (such as resistance to a given antibiotic), which when mixed with a culture of the same type of bacteria that does not carry this trait, can be taken up by a certain fraction of the cells (the "competent" cells) and integrated into their genome. As a result of this process some of the cells acquire the trait carried by the purified DNA. Thus, streptomycin sensitive cells can be transformed to become streptomycin resistant, and all future progeny of these transformed cells will be streptomycin resistant.

TRIPLET STATE:   an excited state in which an orbital electron has an unpaired spin.

TRYPSIN:   one of the proteolytic enzymes that degrade proteins to smaller polypeptide units.

VALENCE ELECTRONS:   the electrons in the outer orbitals of atoms that participate in chemical bonding.

VAN DER WAALS RADII:   the effective atomic or molecular radii of atoms.

VEGETATIVE CELL:   a growing cell.

WILD TYPE:   the original, parental genetic type with which mutants are compared.

ZONE SEDIMENTATION:   the preparative or analytical technique in which material layered on a gradient is resolved into zones of different sedimentation properties in the course of centrifugation of the sample.

# MOLECULAR PHOTOBIOLOGY
*Inactivation and Recovery*

# 1

# Introduction: Basic Principles

## 1-1.  EFFECTS OF ULTRAVIOLET RADIATION UPON BIOLOGICAL SYSTEMS

The first observation of ultraviolet (UV) effects on living systems dates back to 1877 when Downes and Blunt reported that bacteria were inactivated by light. In the years following, the literature on ultraviolet effects grew at an exponentially increasing rate while the understanding of basic mechanisms remained nearly constant. The early work suffered from a lack of appreciation of the necessity for controlling the wavelength of the light as well as from a lack of understanding of the importance of the physiological state of the biological system before, during, and after the irradiation. The next landmark was the finding by Gates in 1928 that the relative effectiveness of killing bacteria by different wavelengths paralleled the absorption spectrum of nucleic acid. The chemical bases for some of the deleterious effects of UV on the nucleic acids did not become evident, however, until the late 1940's. The most recent surge of interest in UV photobiology has been stimulated by the discovery of Beukers and Berends in the early 1960's of UV induced thymine dimers in DNA. As the importance of thymine-dimerization to biological

1

inactivation phenomena became rapidly apparent, there developed a tendency to give it credit for too much of UV photobiology. There are many other types of photoproducts produced in the nucleic acids of cells and certain of these have been isolated and characterized. In some cases, their relative biological importance has also been determined.

We are at the point now where ultraviolet radiation can be used as a very sensitive and specialized probe into the function of the intracellular machinery. Ultraviolet light has many advantages over chemical probes (where permeability of membranes and irrelevant side reactions may complicate analysis) as well as some definite advantages over the use of ionizing radiations (since the latter are not selectively absorbed by a particular class of molecules as is UV). An area in which photobiology has been particularly useful has been in the study of cellular repair mechanisms per se; not merely to learn more about modes of recovery from photochemical damage, but also to understand the broader aspects of the mechanisms that may operate to protect cells from many of the hazards of their natural environments.

In the early evolution of life on earth it is certain that photosynthetic processes were of central importance. They enabled living systems to convert solar energy into chemical energy for the purposes of metabolism and growth; indeed, the sun is still the principal source of energy for the biosphere. However, in general, the effects of absorbed photons on biological molecules are more often destructive and degradative than useful to the necessary functions of these molecules. Mechanisms that afforded protection or recovery from the damaging effects of photons must have developed very early in the evolution of living systems. Later in this text we will outline the various known devices by which currently existing organisms minimize the predominantly deleterious effects of photons that threaten survival.

What are some of the deleterious effects seen at the biological level? We will be considering such effects in a variety of organisms later in the book. However, it should be useful at this point to simply compare the relative sensitivities of a number of the effects at a particular wavelength (2650 Å) of light on the bacterium, *Escherichia coli*, strain B. The most sensitive function to such treatment appears to be the ability of the bacterium to divide. After a dose of only 10 ergs/mm$^2$ less than 10% of the cells are able to divide and most cells are observed to form long "filaments" as cytoplasmic synthesis continues in the absence of division. Somewhat less sensitive is the ability of the irradiated bacterium to form a colony on nutrient agar. Thus at low doses most of the cells eventually regain the ability to divide. A 10% survival of colony-forming ability is seen after doses of the order of 100 ergs/mm$^2$. At such doses one also observes inhibitory effects on the synthesis of DNA, RNA, and protein, although of the three, DNA synthesis is the most sensitive. Production of specific mutations is also observed, but with a frequency considerably less than that for cell killing.

In comparison with the doses indicated above, the sensitivity of bacterial spores to inactivation is strikingly less. A dose of $5 \times 10^4$ ergs/mm$^2$ is required to inactivate dry *Bacillus subtilis* spores to 37% survival. Also, for comparison, the inactivation of the streptomycin marker (a gene) in the *Pneumococcus*-transforming DNA system requires about $10^6$ ergs/mm$^2$ to yield 37% survival, and finally, the inactivation of the enzyme trypsin to 37% survival requires $10^7$ ergs/mm$^2$.

The rationale for using the dose that inactivates to 37% survival for comparing the UV sensitivity of different biological systems is that 0.37 is the value for $e^{-1}$. When inactivation kinetics can be expressed as a simple exponential function of dose, the 37% survival value corresponds to an average of one lethal hit per sensitive target. This is evident from the relationship for the surviving fraction $N/N_0 = e^{-\sigma D}$. In this relationship $N_0$ is the number of sensitive targets present initially, $N$ is the number remaining after a dose, $D$, and $\sigma$ represents the inactivation sensitivity or *cross-section* for the particular target. This analysis will be treated in more detail in Chapter 6.

## 1-2. LAWS OF PHOTOCHEMISTRY

According to quantum theory, radiant energy is transmitted as discrete units called quanta or photons. The energy of a photon is related to the wavelength of light by the Planck relation

$$E = h\nu = hc/\lambda$$

where $E$ is the energy of a single photon in ergs,
  $h$ is Planck's constant, $6.62 \times 10^{-27}$ erg-sec,
  $c$ is the velocity of light, $3 \times 10^{10}$ cm/sec,
  $\nu$ is the frequency of the radiation per sec,
  $\lambda$ is the wavelength of the radiation in centimeters.

The energy of a single photon is so small that in practice one usually refers to the energy of a mole of photons which is called an Einstein (Einstein $= Nh\nu$; where $N$ is Avogadro's number, $6.022 \times 10^{23}$).

We will be concerned with a relatively narrow region of the electromagnetic spectrum (Figure 1-1), principally those wavelengths between 2000 and 3000 Å. These are the important wavelengths for the inactivation of biological systems. We will also consider wavelengths from 3000 to 4500 Å because of their effectiveness in counteracting some of the deleterious effects of ultraviolet light by photoreactivation.

One may wonder why light in this rather narrow range of wavelengths is so effective in killing cells. Wavelengths in the infrared region correspond to energies of 0.01 to 0.1 electron volt (ev), where 1 ev equals 23.08 kcal/mole or $1.602 \times 10^{-12}$ ergs. These energies, when absorbed, result in molecular rotations (rotation of the whole molecule about some axis) and molecular vibrations (the stretching or bending of bonds resulting in displacements of atomic nuclei relative to each other but not affecting equilibrium positions of nuclei). Thus, infrared radiation would not be expected to cause chemical changes in molecules although reaction rates might be increased due to its heating effect. At the other extreme, at wavelengths below about 2000 Å (6.25 ev), the energies are well above the normal range of chemical bond energies. The predominant chemical effects at these energies are random ionization and bond breakage. In Chapter 10 we will be concerned with some of the similarities and differences at the biological and chemical levels between ionizing radiation (X-rays) and ultraviolet light. Energies in the range of 4–6 ev, however, are selectively absorbed by the nucleic acids and proteins and can result in the excitation of valence electrons within these molecules to higher energy levels. Excited molecules show a great probability of undergoing chemical reaction. At the molecular level, it is this photon-induced chemistry that eventually gives rise to an observable effect at the biological level.

The *first law of photochemistry* (Grotthus-Draper Law) states

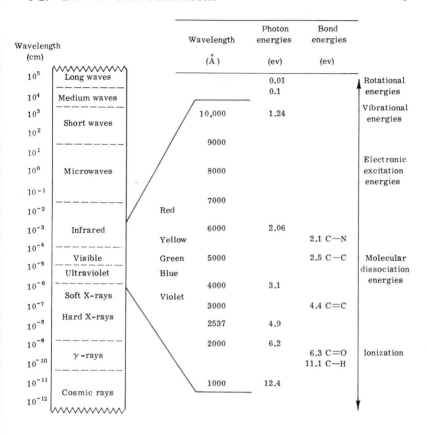

FIGURE 1-1.   The electromagnetic spectrum.

that light must be absorbed by a molecule before photochemistry can occur. If a sample absorbs no light, there can be no photochemical effect. The *second law of photochemistry* (Stark-Einstein Law) states that light absorbed need not necessarily result in photochemistry but if it does then only one photon is required for each molecule affected. The validity of the "one-photon-absorbed-per-molecule-altered" aspect of this second law depends upon the relatively short lifetime of the electronically excited molecule and the relatively low concentration of incident photons in most conventional systems. In recent years, however, biphotonic excitation has been observed in systems using flash photolysis or the laser (see below). These are not violations of the Stark-Einstein Law and in the case of triplet-triplet absorption (see Chapter 3) are to be expected from the very high intensities of light used and the relatively long lifetime of the triplet state.

The Stark-Einstein Law suggests that, in general, a chemical change will *not* be induced by every quantum absorbed by a molecule. The efficiency of producing a photochemical reaction (the *quantum yield*), $\Phi$, is an important concept. It is defined as

$$\Phi = \frac{\text{Number of molecules reacting chemically}}{\text{Number of photons absorbed}}$$

or

$$\Phi = \frac{\text{Number of moles reacting chemically}}{\text{Number of Einsteins absorbed}}$$

Quantum yields may have values as low as $10^{-6}$ for inefficient processes in macromolecular systems and values as high as $10^6$ for photochemically initiated chain reactions (e.g., gas phase photochlorinations). The various mechanisms by which the energy of absorbed photons can be dissipated without chemical consequences will be discussed in Chapter 3.

The chemical data available on the UV inactivation of DNA (and RNA) suggest that the radiation damage is highly localized. The low quantum yields for the inactivation of the nucleic acids ($10^{-3}$ to $10^{-4}$) also support this idea. The quantum yield depends inversely upon the molecular weight and if only a small part of the molecule is altered by the radiation, then the quantum yield does not correctly represent the sensitivity of the DNA. Since the 99% killing dose of *E. coli* B/r (1800 ergs/mm²) dimerizes only 0.1% of the total thymine (25% of the bases in *E. coli* DNA are thymine) and since at least 50% of the inactivating action of UV on transforming DNA can be accounted for by thymine dimerization, we can calculate that only about 0.025% of the molecule of DNA has been photochemically altered by a dose of UV that kills 99% of the bacterial population. If we use this value to correct the molecular weight of the "active center" of the DNA, the quantum yield becomes approximately one, a value much higher than that for any of the free nucleotides and a value more in keeping with the rate of biological inactivation. Clearly the term, quantum yield, is not adequate for describing the UV sensitivity of large polymers when the photochemical modification of only a few of its monomeric units is sufficient to destroy the biological activity of the total molecule. A method to handle this problem for proteins is described in Section 5-4.

In certain situations the *dose rate* may be more important than the total dose. If a dose rate of 1 erg/minute for 10 minutes does not give the same effect as 10 ergs/minute for 1 minute, then *reciprocity* does not hold (Bunsen-Roscoe Law). The reciprocity test may be an important consideration, especially where different wavelengths are compared (as in an action spectrum, see Section 1-4) since for a given light source, the intensities are generally different for different wavelengths. For the inactivation of some enzymes reciprocity holds over a range of $10^5$ in intensity. It may not hold, however, for rapidly growing cells which might divide several times during the irradiation (in an extreme case) or in systems where a slow dark reaction must follow absorption of the photon before an effect can be observed. Reciprocity would also not hold where two simultaneously excited molecules are required for a reaction to occur.

## 1-3.  ABSORPTION SPECTRA

The absorption of light by a molecule results in the conversion of radiant energy into the energy of rotation, vibration and altered electronic configuration within the molecule. The *ground state* of a molecule is that state in which the energies of these internal motions have their minimum values. Other energy states of the molecule are called excited states, and the transition of a ground state molecule to an excited state is called *excitation* (Figure 1-2).

The capacity of a molecule to absorb light of a particular wavelength is dependent both upon the electronic configuration of the molecule and upon the electronic configurations of the possible higher energy states of the molecule. An absorption spectrum of any substance is a quantitative description of the absorptive capacity of the molecule over some range of electromagnetic frequencies and is therefore intimately related to the detailed molecular structure of that substance.

Any absorption spectrum, however complex, may be regarded as a summation of a set of individual absorption bands, each corre-

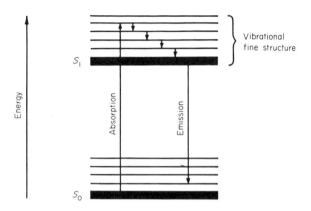

FIGURE 1-2.  Electron excitation by photon absorption. An orbital electron in the
ground state ($S_0$) has been raised to the excited state ($S_1$) and additional energy has
gone into vibrational excitation of the molecule. Following deexcitation of the
vibrations (by molecular collisions) a photon is reemitted as fluorescence when the
electron returns to the ground state ($S_0$).

sponding to a transition between two particular electronic states
plus vibrational states. An individual absorption band may be described by three parameters: (1) the position in the electromagnetic
spectrum, (2) the breadth of the electromagnetic spectrum occupied
by the band, and (3) the intensity of the absorption.

The position of an absorption band in the electromagnetic spectrum is dependent upon the energy difference between the ground
state and the excited electronic configurations since this energy
difference must be supplied by the absorbed photon. These positions
are expressed in terms of wavelength in angstroms or nanometers
(for definitions, see Table 1-1).

The width of an individual absorption band is inversely dependent
upon the duration of the excited electronic state. The width of the
absorption band of a polyatomic molecule (which is the sum of
individual absorption bands) also depends upon the random distribution of the molecules in various vibrational and rotational states
and upon external factors such as molecular aggregation, solvent
effects, etc. (see below). The width of an absorption band is defined
as the spectral separation between the points of half-maximal absorption. This separation is expressed in the units used to express
the position of the absorption maximum.

The intensity of an absorption band (integrated absorption over the band) is determined by the probability that the particular transition will take place when a photon of the appropriate energy comes along and by the duration of the excited state. The intensity of the absorption is usually expressed in terms of optical density or absorbance.

The transmittance, $T$, of a solution is defined as the ratio of the intensity of the light emerging from the solution, $I$, to the intensity of the light incident on the solution, $I_0$, or

$$T = \frac{I}{I_0}$$

The transmittance of a solution containing light-absorbing material depends upon (a) the nature of the substance, (b) the wavelength of light, and (c) the amount of light-absorbing material in the light path. The latter is both a function of the concentration of the substance and the thickness of the solution through which the light passes. The equation relating these factors is known as Beer's Law.

$$\frac{I}{I_0} = 10^{-Elc} \quad \text{or} \quad OD = Elc$$

where $I_0$ = incident light intensity falling on the solution
$\quad I$ = light intensity transmitted through the solution
$\quad c$ = concentration of solute in moles/liter
$\quad l$ = thickness of the solution in cm
$\quad OD$ = optical density = $-\log_{10} I/I_0 = \log_{10} I_0/I$
$\quad E$ = molar extinction coefficient (liters/mole-cm); this is the optical density (under standard conditions of pH, temperature, wavelength, etc.) of a $1M$ solution of a given compound

Recent publications (see *Analytical Chemistry* for recommended nomenclature in spectrometry) will substitute the term absorbance,

TABLE 1-1 WAVELENGTH UNITS

| Unit | Symbol | Length (meters) |
|---|---|---|
| Micrometer | $\mu$m | $10^{-6}$ |
| Micron (obsolete) | $\mu$ | |
| Nanometers | nm | $10^{-9}$ |
| Millimicron (obsolete) | m$\mu$ | |
| Angstrom | Å | $10^{-10}$ |

*A*, for optical density, and molar absorptivity, $\epsilon$, for molar extinction coefficient. Beer's Law then becomes:

$$A = \epsilon lc$$

It is apparent that if the molar absorptivity is known for a given compound, the concentration of a solution of this material can be determined using this formula after first determining the absorbance of the test solution.

Another form of Beer's Law is

$$\frac{I}{I_0} = e^{-nsx}$$

where $x$ = thickness of the absorbing solution

$n$ = number of molecules per unit volume of solution

$s$ = absorption cross-sectional area of a molecule (the product of the probability that a photon passing through a molecule will be absorbed and the average cross-sectional area of the molecule). The absorption cross section is related to molar absorptivity, $\epsilon$, by $s = 3.8 \times 10^{-21}\epsilon$

One sometimes sees statements to the effect that a system does not obey Beer's Law. This is not failure of the law, but rather the failure of the optical or chemical system to conform to the requirements of the relation which are quite explicit, as follows: (a) monochromatic light; (b) independent absorbers, no aggregation with high concentration; and (c) randomly oriented absorbers.

The empirical correlation of the spectral position of the absorption bands in the ultraviolet with certain chemical structures was begun about 1885 and has been expanded as improvements in technique have simplified the task of measuring absorption spectra. These correlations indicate that the absorption bands of compounds composed exclusively of saturated linkages occur usually below 2000 Å. Because many spectrophotometers do not record below 2000–2200 Å, these short wavelength bands of saturated compounds, rising in absorption with decreasing wavelength, are often referred to as "end" absorption.

Compounds with single unsaturated bonds such as $C = C$, $C = O$, or $C = N$, have absorption bands, usually weak, in the region of 1900–3000 Å, the actual wavelength being dependent upon the adjacent parts of the molecule.

Strong absorption bands in the region 2000–4000 Å are always correlated with molecular structures containing conjugated double bonds: in general, the larger the conjugated structure, the stronger is the absorption and the longer the wavelength of the maximum absorption. Ring structures with conjugated double bonds often exhibit a particularly strong absorption and the absorption of such compounds can be markedly affected by the addition of side chains or auxiliary groups, especially if the latter are charged (i.e., $-NH_2$ and $-OH$). Such groups may affect the spectral position, structure, and intensity of an absorption band of a conjugated system (Figure 1-3). In general, the variation of any factor that influences the elec-

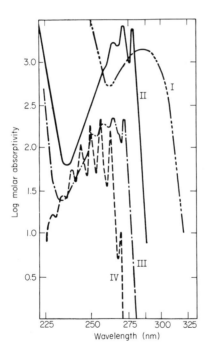

FIGURE 1-3. Ultraviolet absorption spectra of some derivatives of benzene. I, aniline; II, phenol; III, chlorobenzene; IV, benzene; all in heptane. [Adapted from K. L. Wolf and W. Herold, *Z. Physik. Chem.* **B13**, 201 (1931), and modified from R. L. Sinsheimer, *in* "Radiation Biology" (A. Hollaender, ed.), Vol. 2, p. 171. McGraw-Hill, New York, 1955.]

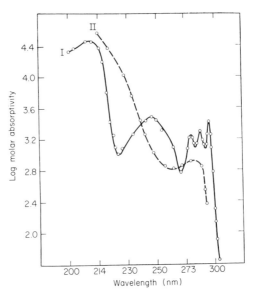

FIGURE 1-4.    The absorption spectrum of phthalic anhydride. I, in hexane; II, in alcohol. [Modified from S. Menczel, *Z. Physik. Chem.* **A125**, 161 (1927).]

tronic configuration of the absorbing molecules will affect the absorption spectra.

In addition, certain factors may alter the technical conditions of the absorption measurements and thereby affect the spectrum. The choice of solvent can influence the position, width, and intensity of the absorption bands. Absorption bands are widened in polar solvents in part because of increased molecular interaction (Figure 1-4).

In aqueous solutions of substances containing dissociable groups, the pH of the solution will usually have a marked effect on the absorption spectrum. Ionization of any such group, resulting in gain or loss of charge, will certainly alter the basic electronic configuration of the molecule and thereby the spectral distribution of absorption (Figure 1-5).

In very concentrated solutions, the association of solute molecules may cause a modification of their absorption spectrum. This effect may give rise to a nonlinear relation between the absorbance of such solutions and the solute concentration (Figure 1-6).

The temperature of an absorbing substance significantly affects

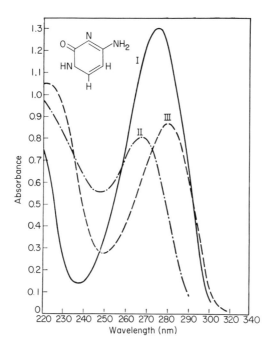

FIGURE 1-5.  The absorption spectrum of cytosine at different values of pH. I, pH 1.2; II, pH 6.0; III, pH 12.7. [Adapted from J. F. Scott, unpublished data (1951), cited by R. L. Sinsheimer, *in* "Radiation Biology" (A. Hollaender, ed.), Vol. 2, p. 181. McGraw-Hill, New York, 1955.]

its absorption spectrum by influencing the statistical distribution of molecules among various vibrational states associated with the lower energy electronic state and by influencing the velocity of Brownian motion which in turn determines the frequency of molecular collision. The latter influences the duration of the excited state and the extent of the distortion of the molecular electronic configurations by the electromagnetic fields of neighboring molecules (Figure 1-7).

The energy of electronic excitation may be rapidly dissipated by conversion to vibrational energy under conditions of appreciable intermolecular contact. Such dissipative effects are in part responsible for the broadening of absorption bands of substances in solution as compared to their vapor absorption spectra.

The absorption of conjugated bond groups separated within a

given molecule by two or more saturated bonds is usually independent and simply additive. In large polar macromolecules such as
proteins and nucleic acids, however, the ultraviolet absorption
spectrum of the polymer is often not strictly the linear sum of the
absorption of its component conjugated groups (Figure 1-8). This
nonadditivity of absorption is often referred to as *hypochromicity*
or *hyperchromicity*. The two terms refer to the same principle, but
there is a slight difference in the numbers assigned to each. If the
absorbance of a given oligonucleotide is lower than that of its
constituent mononucleotides, it is hypochromatic or exhibits hypochromicity. If the oligonucleotide is hydrolyzed to mononucleotides
with the resulting increase in absorbance, it exhibits hyperchromicity. If a given oligonucleotide exhibits 75% of the absorption
of its component mononucleotides, its hypochromicity is 25/100
or 25%, whereas its hyperchromicity is 25/75 or 33%. Hypochromicity can largely be explained by the coulombic interaction of the
ordered bases in the polymer and by the extent of the interaction of
the bases with the solvent (see further discussion in Section 3-6).

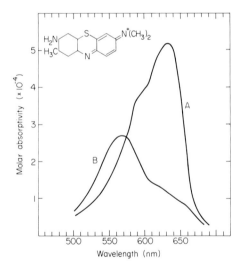

FIGURE 1-6.    The absorption spectrum of toluidine blue in water at different concentrations. A, $5.17 \times 10^{-7}$ *M*; B, $3.88 \times 10^{-4}$ *M*. [Adapted from L. Michaelis,
*Cold Spring Harbor Symp. Quant. Biol.* **12**, 131 (1947).]

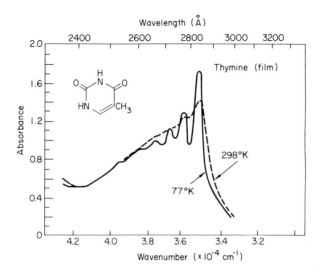

FIGURE 1-7. The absorption spectrum of a thin film of thymine at room temperature and at liquid nitrogen temperature. [Adapted from R. L. Sinsheimer, J. F. Scott, and J. R. Loofbourow, *J. Biol. Chem.* **187**, 313 (1950).]

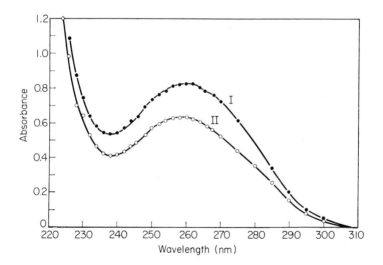

FIGURE 1-8. Change in absorption spectrum of DNA upon digestion with deoxyribonuclease. I, digested; II, undigested. [M. Kunitz, *J. Gen. Physiol.* **33**, 349 (1950).]

1-4.  ACTION SPECTRA

The reason that only certain wavelengths of light are absorbed by a given type of molecule is that the energy of an absorbed photon must exactly equal the excitation energy for some allowed excited state of the molecule. For polyatomic molecules, the transition to an excited electronic configuration is characterized by a fairly broad absorption band, in which varying degrees of vibrational and rotational excitation are superimposed upon a fundamental electronic excitation. The molecular groupings giving rise to this absorption in proteins are principally the aromatic amino acids, resulting in absorption maximum around 2800 Å. For the nucleic acids, the chromophores responsible for the absorption maximum around 2600 Å are the heterocyclic bases, the purines and pyrimidines. The absorption spectrum thus provides a suggestive pattern in choosing the wavelengths for photochemical and photobiological investigations (Figure 1-9).

In many biological processes, however, a photobiological phenomenon is observed before the definite identity of the active absorbing molecule is known. In such a case it is very useful to measure the wavelength dependence of the phenomenon. The plot of the quantitative biological (or chemical) response as a function of wavelength is called the *action spectrum*. The action spectrum can often give a clue to the identity of the chromophore absorbing the radiant energy. It was found, for example, that a plot of the effectiveness of the various wavelengths of ultraviolet light in killing bacteria showed a maximum around 2650 Å and the shape of the curve was very much like that of the absorption spectrum of the nucleic acids (Figure 1-10). It was thus assumed that for the killing of bacteria the primary target of the ultraviolet radiation was DNA. In view of our current knowledge concerning the role of DNA in mutations, it is not at all surprising to find that the action spectrum for mutation production is also similar to the absorption spectrum of DNA (Figure 1-11).

Not all action spectra are similar to the absorption spectrum of DNA. The immobilization of *Paramecia* yields an action spectrum similar to the absorption spectrum of albumin, thus suggesting that the sensitive target in the cilia of the *Paramecia* is protein (Figure 1-12).

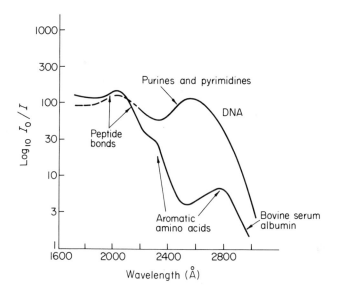

FIGURE 1-9.   Ultraviolet absorption spectra of 1% solutions of a typical protein (bovine serum albumin) and of DNA at pH 7. [Adapted from R. B. Setlow and E. C. Pollard, "Molecular Biophysics," p. 226. Addison-Wesley, Reading, Massachusetts, 1962.]

For the retardation of cleavage in the sea urchin, the action spectrum  for irradiated sperm resembles the absorption spectrum of DNA, having a peak at about 2600 Å; while that for the egg resembles protein and has a peak at about 2800 Å (Figure 1-13).

The activation of sea-urchin eggs to undergo parthenogenetic development has an action spectrum similar to the absorption spectra of compounds composed primarily of saturated linkages. This action spectrum does little toward the identification of the active chromophore but is important in eliminating DNA or proteins (containing a significant amount of aromatic amino acids) as the active chromophore (Figure 1-14).

There are numerous other examples of action spectra in the literature (for photoreactivation, photosynthesis, phototropism, phototaxis, etc.). Even though action spectra may frequently provide useful information, they are often not as easy to interpret as in some of the examples cited above.

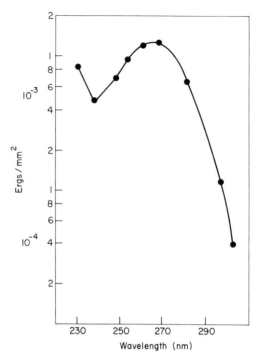

FIGURE 1-10.   Action spectrum for bacterial killing. Curve of the reciprocals of the incident energies required for 50% killing of *E. coli* versus wavelength. [Adapted from F. L. Gates, *J. Gen. Physiol.* **14**, 31 (1930).]

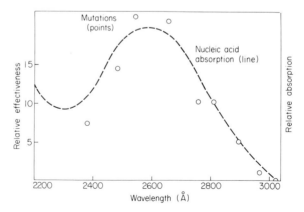

FIGURE 1-11.   Action spectrum for mutation production in corn. [Adapted from L. J. Stadler and F. M. Uber, *Genetics* **27**, 84 (1942).]

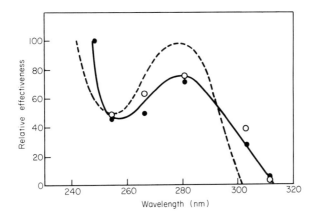

FIGURE 1-12. Action spectrum for immobilization of paramecia (——). O denotes fed, and ● denotes starved paramecia. -----, absorption spectrum of albumin. [Adapted from A. C. Giese, *Physiol. Zool.* **18**, 223 (1945).]

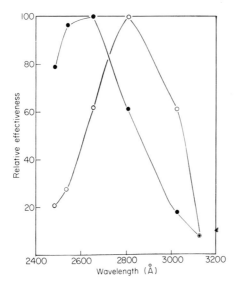

FIGURE 1-13. Action spectra for retardation of cleavage in sea urchin for irradiated sperm ● and irradiated eggs O. [Adapted from A. C. Giese, *Quart. Rev. Biol.* **22**, 253 (1947).]

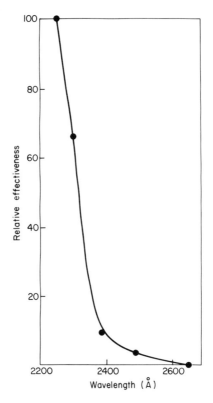

FIGURE 1-14.   Action spectrum for activation of *Arbacia* eggs to undergo parthenogenetic development. [Adapted from A. Hollaender, *Biol. Bull.* **75**, 248 (1938).]

GENERAL  REFERENCES

*Absorption Spectra*
W. Heitler, "The Quantum Theory of Radiation," 2nd ed., Oxford Univ. Press,
    London and New York, 1944.
R. L. Sinsheimer, Ultraviolet absorption spectra. *In* "Radiation Biology (A. Hollaen-
    der, ed.), Vol. 2, p. 165. McGraw-Hill, New York, 1955.
C. F. Hiskey, Absorption spectroscopy. *Phys. Tech. Biol. Res.* **1**, 73 (1955).
J. F. Scott, Ultraviolet absorption spectrophotometry. *Phys. Tech. Biol. Res.* **1**,
    131 (1955).
H. H. Jaffe and M. Orchin, "Theory and Applications of Ultraviolet Spectroscopy."
    Wiley, New York, 1962.

*Action Spectra*

A. C. Giese, Ultraviolet radiations and life. *Physiol. Zool.* 18, 233 (1945).

A. C. Giese, Studies on ultraviolet radiation action upon animal cells. *In* "Photophysiology (A. C. Giese, ed.), Vol. 2, p. 203. Academic Press, New York, 1964.

R. B. Setlow, Action spectroscopy. *Advan. Biol. Med. Phys.* 5, 37 (1957).

R. B. Setlow, Ultraviolet wavelength-dependent effects on proteins and nucleic acids. *Radiation Res.* Suppl. 2, 276 (1960).

P. Halldal, Ultraviolet action spectra in algology. *Photochem. Photobiol.* 6, 455 (1967).

# 2

# Experimental Procedures

## 2-1.  LIGHT SOURCES

What considerations are important in selecting a lamp for photo-chemical studies? Certainly there are the practical considerations of simplicity of construction and operation, useful life, availability, and cost. The more important considerations, however, have to do with the intensity and the spectral output of the light (remember that light must be absorbed before photochemistry can occur). Some lamps give a nearly continuous spectrum of energies while other lamps emit light principally at one or more wavelengths. The best known of the latter type is the low pressure mercury vapor lamp which, if encased in silica glass, emits almost all (i.e. > 95%) of its radiant energy at 2537 Å. Since this is very near the wave-length for maximum efficiency of bacterial killing (Figure 1-10), this lamp is known as a germicidal lamp. If the lamp casing is quartz, however, a significant amount of ozone producing radiation at 1849 Å will be transmitted but this can be removed by using a filter composed of acetic acid or ethanol that will absorb wave-lengths below about 2300 Å. High pressure mercury vapor lamps

23

provide a broad spectrum of intense ultraviolet and visible radiation but tend to have their greatest intensity at the longer wavelengths. At lower intensities, but also giving a broad spectrum of emission, are the tungsten lamps and fluorescent lamps. These lamps emit primarily in the visible region but the latter may also emit a non-negligible amount of energy in the near ultraviolet. The tungsten and fluorescent lamps are mainly used to provide energy for the photoreactivation of ultraviolet induced damage in biological systems (see Section 7-3), and for studying photodynamic action (the inactivation of a biological system when exposed to the combination of visible light, a dye and oxygen; Chapter 9).

## 2-2. WAVELENGTH SELECTION

If one wished to study the biological and chemical effects of radiation at 2537 Å, then the low pressure mercury discharge lamp would be ideal, (providing that the intensity was adequate; see below) and little or no filtration of extraneous light would be necessary. With the continuous or broad spectrum lamps, however, one must filter out those wavelengths that are not of interest (or that might cause side effects that would complicate the analysis of the system under study). The simplest filters are absorption filters. By virtue of the absorption spectra of their components these filters absorb, more or less strongly, certain wavelengths while transmitting other wavelength regions. Such filters are simple to use and may be made in large dimensions. On the other hand, it is difficult to obtain absorption filters which can provide both a narrow transmission band and high transmission within the band. Absorption filters may be made of glass, (Figure 2-1) or liquid- or gas-filled cells, or combinations of these. If the absorption of the ultraviolet radiation involves a photochemical decomposition of some component of the filter, the filter may have to be renewed frequently.

Commercially available *interference filters* can also be used to help localize different regions of the spectrum. Conventional types consist of two semitransparent evaporated metal films on glass or quartz plates which are separated by a transparent layer of a thickness comparable to the wavelength of light to be passed. The filter reflects all incident light except for the chosen wavelengths. Under favorable optical conditions these filters can provide a peak trans-

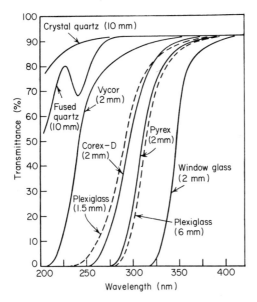

FIGURE 2-1.   Spectral transmission of various types of transparent materials. [Adapted from R. B. Withrow and A. P. Withrow, *in* "Radiation Biology" (A. Hollaender, ed.), Vol. 3, p. 125. McGraw-Hill, New York, 1956.]

mission of about 35% with a band width of about 100 Å. These filters are primarily useful in the visible region, but some interference filters are now available for the ultraviolet region.

A more elegant method for obtaining a particular wavelength is to disperse the various radiant energy lines and then selectively pick out the wavelength of interest. Such an instrument is called a *monochromator*. There are two main methods for dispersing light. One method uses a prism (Figure 2-2). The parallel rays of white light are separated into their component colored parts due to the fact that light of different wavelengths is refracted to a different extent when passing from one medium to another (air to prism to air). If one is working only with visible light then the prisms can be made of glass, but if ultraviolet light is required the prisms must be made of quartz.

A second technique for obtaining monochromatic light is to use a *diffraction grating*. The grating consists of a number of thin, parallel, reflecting strips and functions by breaking up a wave front of radiation into a number of narrow, parallel zones evenly spaced by appropriate distances; the waves propagating from each zone will interfere with those from all other zones so as to produce a diffraction pattern. For any given wavelength, there will be some direction or directions in which the waves from each zone will be in phase to produce a maximum of intensity. In another direction, waves of the given wavelength will largely cancel each other. By thus deviating radiation of different wavelengths into different angles, a grating can serve as a dispersing element.

On rainy days, the clouds sometimes act as dispersing elements and separate the sunlight into its component parts, resulting in the appearance of a rainbow. In the laboratory, the rainbow produced by passing light through a prism can be positioned by rotating the prism such that light of a chosen wavelength will pass through the exit slit of the monochromator and thus expose the sample to be irradiated (Figure 2-2). The problem with any monochromator to be used in photochemical work is primarily that of obtaining adequate intensity, although spectral purity is also a very major problem. Placing a high intensity lamp on a small monochromator usually results in a loss of spectral purity due to an increase of stray light. High intensity light is also detrimental to the lifetime of replica diffraction gratings.

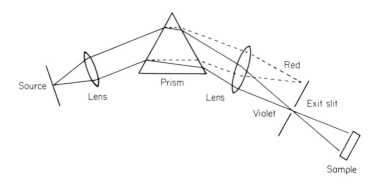

FIGURE 2-2. Schematic of a single prism monochromator. The parallel rays of the white light are separated into their component colored parts due to the fact that light of different wavelengths is refracted to a different extent when passing from one medium to another (air to prism to air). See text for further discussion. (Adapted from J. Jagger, "Introduction to Research in Ultraviolet Photobiology." Prentice-Hall, Englewood Cliffs, New Jersey, 1967.)

Another apparatus that yields intense monochromatic light and which is beginning to be used in photobiology and photochemistry is the *laser*. The term is an acronym for "*l*ight *a*mplfication by *s*timulated *e*mission of *r*adiation." The first solid-state laser was made from a ruby crystal about four centimeters long and half a centimeter across. The ends of the ruby are polished optically flat and parallel and one end is heavily silvered while the other end is only slightly silvered. A powerful source of light (an electronic flash tube) is coiled around the ruby crystal. When this tube is flashed, the majority of the chromium atoms in the ruby crystal are raised to a metastable excited state. When these atoms return to the ground state they give off photons at 6943 Å. If a ground state atom absorbs one of these photons it can be raised to a metastable state, or if the photon strikes an atom already in the metastable state it will cause it to give off instantly a photon at 6943 Å. Any photons directed at angles to the long axis will be reflected back and forth by the silvered ends of the ruby causing more and more photons to be liberated, and finally, when the intensity reaches a certain value, it will penetrate the lightly silvered end of the ruby giving rise to an intense pulse of coherent (all waves in phase) monochromatic light.

Continuous lasers are currently available with an output in the visible region of approximately 1 watt (see Table 2-1 for energy units). Pulsed lasers have achieved power in the megawatt range. When the power of 1 watt is diffused from its normal focused area of about $2 \times 2$ mm to 1 $cm^2$ the power would be $6 \times 10^8$ ergs/$cm^2$/min. Spreading out the beam further would, of course, reduce the average intensity per unit area. For comparison, the amount of power recovered from a 500 watt high pressure mercury lamp at 366 nm (the peak output between 220–550 nm) after passing through a diffraction grating monochromator is about $10^6$ ergs/$cm^2$/min over an area of about $1 \times 3$ cm. At 220 nm the output drops to about $10^4$ ergs/$cm^2$/min. An unfiltered 15 watt germicidal lamp has an output (at 253.7 nm) of about $10^5$ ergs/$cm^2$/min at a distance of 43 cm and over an area of about $15 \times 35$ cm. The intensity of the light falling upon a sample can be increased by moving the sample closer to this lamp (see inverse-square law; Section 2-3).

Lasers are currently available (both solid, liquid and gas lasers) that emit wavelengths as low as 350 nm, and others are being developed to emit in the short-wave ultraviolet region. Even if lasers of suitable power and wavelength were available, the cost

TABLE 2-1. UNITS OF SPECTRAL ENERGY AND POWER

| | |
|---|---|
| 1 erg | $= 6.242 \times 10^{11}$ electron volts |
| | $= 2.389 \times 10^{-8}$ gram calories |
| 1 joule | $= 10^7$ ergs |
| 1 watt | $= 1$ joule/sec $= 10^7$ ergs/sec |
| 1 quantum $= h\nu$ | $= \dfrac{hc}{\lambda} = \dfrac{1.986 \times 10^{-8}}{\lambda \text{ (Angstroms)}}$ ergs |
| 1 Einstein $= Nh\nu$ | $= \dfrac{1.196 \times 10^{16}}{\lambda \text{ (Angstroms)}}$ erg |

(approximately \$10,000) will probably prevent their widespread use in photobiology for some time to come.

## 2-3. MEASUREMENT OF ENERGY OUTPUT

Once one has obtained a light of adequate intensity and has been satisfied as to its spectral purity, it then remains to determine the total amount of radiant energy in absolute units being delivered by the source (Table 2-1). One useful detector is the *thermopile*, an instrument that responds to the total energy of the radiation *independent of wavelength*. This instrument consists of a series of junctions of two dissimilar metals. Alternating junctions are soldered to a thin metal foil which is later blackened with lampblack to insure that it will absorb *all* incident radiation. The thermoelectric potential between these hot junctions and the junctions not exposed to the radiation can be measured with a sensitive galvanometer, the deflection of which is proportional to the intensity of the incident radiation. A thermopile is usually calibrated by measuring its response to light from a standard lamp (available from the U.S. Bureau of Standards) operated under precisely defined conditions of voltage and geometry. The thermopile galvanometer readings can then be converted to ergs per second.

Although the thermopile is considered a primary detector for the energy calibration of monochromatic light, its main disadvantages are its sluggish response (it may take several minutes to reach equilibrium) and its troublesome response to any other heat sources in the room (e.g., the experimenter). In general it is more convenient to use the thermopile to calibrate the response of a photoelectric cell (considered a secondary detector for the calibration of light sources). Although photoelectric cells have a very fast response time they do not respond with equal sensitivity to all wavelengths (as does a thermopile) so care must be used in choosing a photo-electric cell with sufficient sensitivity for the wavelengths to be used and it must then be calibrated at each wavelength.

Still another method used for determining the intensity of the light source is chemical *actinometry*. There are some photochemical reactions whose quantum yields (defined in Section 1-2) are so well known that an analysis of the extent of the reaction gives an imme-diate measure of the integrated amount of energy falling on the reaction cell. Actinometers are not primary detectors but can be calibrated against a thermopile. The advantages of actinometers are that (1) no electrical measurements need be made, and (2) a vessel of any size or shape can be adapted to hold the actinometer reactants. The most widely used actinometric reaction, but certainly not the only one available, is the photosensitized decomposition of oxalic acid by uranyl ion. The extent of photolysis of the oxalic acid is determined by titration with $KMnO_4$. The quantum yield for this reaction has been measured over the range of 2540 to 4350 Å (see E.J. Bowen, 1946).

One can adjust the exposure dose rate to a sample by moving the sample to be irradiated either closer to or further away from the light source, taking into account the *inverse-square law*. The law states that the intensity of light from a point source varies inversely as the square of the distance from the source. For actual sources of finite size, the inverse-square law can be applied with an error of less than 1% when the largest dimension of the source or receiver is not more than one-tenth the distance between the two. The error increases as the sample to be irradiated is brought closer to the lamp.

The opposite extreme from a point source is a uniformly distrib-uted source such as a large bank of fluorescent lamps with closely spaced tubes. For infinite, uniformly distributed sources, the in-tensity is independent of the distance. (This is equally true for a laser beam). Lamps mounted in reflectors have properties that are intermediate between those of a point source and those of a uni-formly distributed source.

2-4. GENERAL PROBLEMS IN DOSIMETRY

The lack of agreement on the photobiological sensitivities of biological systems that is sometimes evident when the results of different laboratories are compared can often be explained by dosimetry considerations. There are a number of highly accurate methods for determining the incident intensity and wavelengths of a photon flux, as described in the previous section. However, there can be a number of reasons why an irradiated sample may not receive the measured incident intensity. The most obvious reason would be the case in which the sample absorbs an appreciable fraction of the incident photons. To take a simple example, consider the irradiation of a culture of bacteria in a standard 1 cm quartz cuvette. At time $t_0$, a bacterium near the front wall of the cuvette (toward the source) may receive nearly the incident photon flux. (There may be some loss of the incident flux by reflection from the front wall of the cuvette but this can be determined by absorption measurements on the cuvette in the absence of sample.) A bacterium near the back of the cuvette may receive considerably less than the incident photon flux.

At a concentration of $2.5 \times 10^8$ bacteria per milliliter there is only 31% transmittance of 260 nm photons through a 1-cm path length cell. It is immediately evident that the *average* dose received by a bacterium in this cuvette, even with stirring, will be less than the *incident* dose. An obvious and simple solution is to use optical path lengths that are short enough so that the emergent intensity differs very little from the incident intensity. In some situations, however, the reduction of sample concentration or path length will not leave enough sample for subsequent analytical purposes.

The alternative is to determine the actual average intensity from a measurement of the absorbance of the sample. The average intensity $\bar{I}$ is really the summation of all volume elements $dx$ and the actual intensities at their positions divided by the number of such elements. A calculation by Morowitz based upon these considerations yields an expression of the form:

$$\bar{I} = I_0 \, (1 - e^{-nsx})/nsx.$$

From this expression (the terms have been defined in Section 1-3) it can be shown that the correction factor for the dose received by a culture of bacteria with a transmittance of 31% in a 1-cm path length is 0.6. Thus, the average incident dose to the bacteria will be only 60% of the incident dose to the front of the cuvette. Table 2-2 gives the correction factors for different sample transmittances.

Although the mathematical treatment of this problem indicates that the average dose is little effected by stirring, in practice however, stirring during irradiation can be shown to be very important. Cells that might stick temporarily to the back wall of the cuvette would receive much less radiation than the average dose. The resultant dose-effect curve for survival would be biphasic and would therefore incorrectly suggest the presence of a radiation resistant fraction of cells in the population. The interpretation of survival curves will be discussed more fully in Chapter 6.

Another problem in the determination of the average dose may be due to tendencies of the sample units to aggregate. You will remember that Beer's Law requires that the sample consist of independent and randomly oriented absorbers. The aggregation of the absorbers would drastically affect the absorbance (Figure 1-6) and hence the calculated average dose.

Finally, we should interject a note of caution in that the measured emergent intensity from a sample does not indicate merely the absorbed radiation but also the scattered radiation. Associated with

TABLE 2-2. THE AVERAGE INCIDENT DOSE ON A SAMPLE AS A FUNCTION OF THE TRANSMITTANCE OF THE SAMPLE

| Transmittance | Absorbance | Correction factor (stirred sample) |
|---|---|---|
| 100 | 0 | 1 |
| 90 | 0.046 | 0.95 |
| 80 | 0.097 | 0.90 |
| 70 | 0.155 | 0.84 |
| 60 | 0.222 | 0.78 |
| 50 | 0.301 | 0.72 |
| 40 | 0.398 | 0.66 |
| 30 | 0.523 | 0.58 |
| 20 | 0.699 | 0.50 |
| 10 | 1.000 | 0.39 |
| 5 | 1.301 | 0.32 |

Adapted from H. J. Morowitz, *Science* 111, 229 (1950).

light is a periodically varying electric field. When light falls upon matter this field induces a polarizability which oscillates with the same frequency as that of the incident light. The material then acts as a secondary source and the light is reemitted in various directions. Rayleigh scattering of light is dependent upon the size of the particle (roughly as the 6th power of the particle radius). A particle 100 times the size of a given particle will scatter as much light as 10,000 of the given particles. Rayleigh scattering also depends upon the reciprocal of the fourth power of the wavelength. This explains why the sky is blue rather than red. An example of the contribution of light scattering to the absorption spectrum of a culture of bacteria is given in Figure 2-3.

## 2-5. FLASH PHOTOLYSIS

Another technique currently yielding useful information in photochemistry is that of flash photolysis. Many photochemical processes consist of a series of dark reaction steps following the primary photochemical reaction. The overall quantum yield may, therefore, give no information about the primary or intermediate processes taking place. Until recently the primary photochemical steps had normally to be inferred from an analysis of the kinetic data for the formation of final products. This is because the primary photochemical reaction intermediates are so short-lived that their actual concentrations under normal experimental conditions are much too low to be observable. If, however, one increases significantly the intensity of the light source a greater yield of products will be produced per unit of time; then by using a spectrograph, fast changes in the absorbance of the irradiated solution can be observed. A spectrograph is simply a monochromator that takes the light transmitted through the sample and separates it out into its component wavelengths. This light is then used to expose a photographic film. The amount of exposure of the photographic film at a given wavelength region is proportional to the intensity of the light transmitted at that wavelength.

For flash photolysis a light whose intensity is about $10^7$ times more intense than that ordinarily used for photochemistry is coiled around a long cylindrical cuvette containing the substance to be

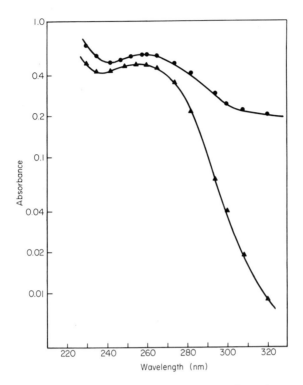

FIGURE 2-3. The effect of light scattering on the apparent absorption spectrum of a culture of bacteria. The sample contained $2.5 \times 10^8$ cells/ml of *E. coli* strain 15T$^-$ and the measured absorbance for whole cells is shown by the upper curve, ●. The bacteria were lysed by the addition of KOH, the solution was neutralized with HCl and the absorbance was again determined, ▲.

studied. This light is flashed for 10 to 100 microseconds and then a spectroscopic light is flashed from a second direction (parallel to the long axis of the cuvette). The transmittance of the spectroscopic light through the irradiated sample is captured on film by the spectrograph and films can then be exposed at intervals from about 20 microseconds after the excitation flash. One can thus follow the appearance and decay of light absorbing photochemical intermediates rather than being restricted to the study of the end product of a photochemical reaction.

GENERAL REFERENCES

E. J. Bowen, "The Chemical Aspects of Light." Oxford Univ. Press (Clarendon), London and New York, 1946.

L. J. Buttolph, Practical applications and sources of ultraviolet energy. *In* "Radiation Biology" (A. Hollaender, ed.), Vol. 2, p. 41. McGraw-Hill, New York, 1955.

J. F. Scott and R. L. Sinsheimer, Technique of study of biological effects of ultraviolet radiation. *In* "Radiation Biology" (A. Hollaender, ed.), Vol. 2, p. 119. McGraw-Hill, New York, 1955.

R. B. Withrow and A. B. Withrow, Generation, control, and measurement of visible and near-visible energy. *In* "Radiation Biology" (A. Hollaender, ed.), Vol. 3, p. 125. McGraw-Hill, New York, 1956.

P. M. B. Walker, Ultraviolet absorption techniques. *Phys. Tech. Biol. Res.,* 3, 401 (1956).

A. D. McLaren and D. Shugar, "Photochemistry of Proteins and Nucleic Acids." Pergamon Press, Oxford, 1964.

L. R. Koller, "Ultraviolet Radiation." Wiley, New York, 1965.

J. G. Calvert and J. N. Pitts, Jr., "Photochemistry." Wiley, New York, 1966.

J. Jagger, "Introduction to Research in Ultraviolet Photobiology." Prentice-Hall, Englewood Cliffs, New Jersey, 1967.

# 3

# Quantum Photochemistry

## 3-1. PROPERTIES OF PHOTONS AND ELECTRONS

In this chapter we will consider what happens at the electronic level when a photon is absorbed by a molecule; why it is absorbed in the first place, and why the absorption event can lead to chemical reaction under the proper circumstances. This should then provide some basis for the understanding of the photon induced changes in nucleic acids and proteins to be discussed later.

The transition from classical physics to atomic physics is intimately linked with the transition from continuous to discrete atomic phenomena. This is true for radiation as well as for matter. Whereas classical physics is concerned with light waves that are continuous in space and time (e.g., a spherical wave emitted by a light source), atomic phenomena such as the interaction between radiation and matter can only be explained by assuming that radiation is emitted and absorbed as single quanta. According to Planck's fundamental relationship ($E = h\nu$), the energy of these quanta (which we call photons) is proportional to the frequency of a radiating dipole. Classically, the origin of all electromagnetic radiations is in the oscillation of electric charges or dipoles. The relationship was discovered by Planck in 1900 when he tried to derive theoretically an observed empirical relation for the spectral energy distribution of blackbody radiation. (A blackbody radiator is one that completely absorbs incident radiation of every wavelength.) Such a radiator emits continuous radiation whose spectral energy distribution depends exclusively upon the absolute temperature. Thus, as illustrated in Figure 3-1, Planck obtained a family of curves relating energy to wavelength at different temperatures. The rigorous theoretical derivation of a formula that explained these curves was *not* possible on the basis of the assumption that the oscillating dipoles could vibrate with *any* amplitude and frequency. The correct relation could be explained only by the revolutionary assumption that the vibrating energy of a resonator was proportional to its frequency and always an integral multiple of $h\nu$. This assumption of quantized states of a blackbody radiator was the origin of quantum mechanics. (Although the purely classical Rayleigh-Jean's theory had previously implied that $\nu$ was quantized.) Some years later Einstein applied Planck's formula to radiation itself and conceived the idea of photons. So, instead of a spherical wave being radiated from a source we have an irregular statistical emission of particles (i.e., photons)

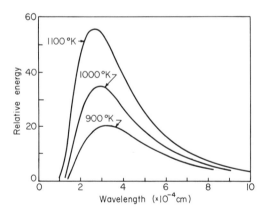

FIGURE 3-1.   The spectral energy distribution of blackbody radiation, showing the dependence upon temperature. The Planck constant, $h$, as well as the Boltzmann constant, $k$, can be calculated from measurements on the temperature dependence of blackbody radiation as shown in this figure. [Adapted from R. W. Ditchburn, "Light," p. 565. Wiley (Interscience), New York, 1953.]

into all directions of space. Of course photons also have wave aspects and these properties can be demonstrated in other experiments. The contradictory characteristics of particles and waves require that one cannot simultaneously apply a particle description and a wave description in the same experiment. The wave model can be used to describe experiments on interference and diffraction while the particle model is necessary to explain certain other phenomena (e.g., photoelectric effect, the Compton effect, pair production, and pair annihilation). The theoretical *principle of complementarity,* formulated by Niels Bohr in 1928, resolves the wave-particle paradox by concluding that it is necessary to attribute both wave characteristics and particle characteristics to electromagnetic radiation and also to material particles such as electrons.

Having discussed the particulate aspect of radiation we should also briefly consider the wave nature of electrons. Interference patterns of electron beams (or even of helium ion beams for that matter) provide quite a convincing demonstration of the wave properties of particles. The most successful theoretical treatment of the motion of electrons in atoms has been that of wave mechanics. In fact the basic wave equation for the hydrogen atom has been solved to give *exact* answers—that is, to express precisely the probability of finding the electron at any given point in space. The solution for this problem comes from a consideration of the motion of an electron in the coulombic field of the positive proton nucleus. For more complex systems exact solutions are not obtainable but approximation methods have been worked out on the assumption that the solutions will resemble in form those for the hydrogen atom.

Solutions to the wave equation for the hydrogen atom are possible only for certain discrete values for the energy of the atom. Thus, the energy that binds the electron to the nucleus is quantized. Solutions of the wave equation that correspond to various discrete energy levels are obtained as functions of the position of the electron with respect to the proton. The wave equation gives the probability that an electron is at a certain point in space. These probability functions are tabulated in order of increasing energy and the lowest energy level is most stable and is called the *ground state.* The wave function of lowest energy for the hydrogen atom is spherically symmetrical around the proton and it is referred to as the 1s orbital (Figure 3-2). The next, more energetic, orbital is also spherically symmetrical and is called the 2s orbital. The next three orbitals have the same

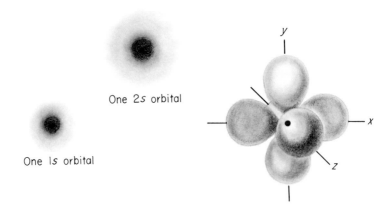

One 2s orbital

One 1s orbital

Three perpendicular 2p orbitals

FIGURE 3-2.    The orbitals for the hydrogen atom.

energy values as the $2s$ orbital, but they are symmetrical, respectively about the three perpendicular axes. They are called the $2p_x$, $2p_y$, and $2p_z$ orbitals, respectively. Each orbital has three quantum numbers (e.g., Cartesian coordinates) that describe the positioning of the electron in space. To complete the description of the state of the electron we need a fourth quantum number called the *spin*. In addition to its orbital motion about the nucleus the electron can be considered to rotate about its own axis. Since rotation of a charged sphere about its axis is equivalent to a circular current, it gives rise to a magnetic field whose direction is in that of the axis. This may be positive or negative, depending upon which way the electron is spinning. Finally, in addition to the quantization of physical parameters a further restriction must be imposed that is known as the *Pauli exclusion principle*. It specifies that no two electrons in an atom can be in the same detailed state (i.e., no two electrons can have the same four quantum numbers). Therefore, two electrons might occupy the same orbital but their spins would have to be in opposite directions. Also no more than two electrons can occupy the same orbital.

## 3-2. COVALENT BONDS

So far we have been considering only a single atom. Chemical

bonding between atoms occurs because a single electron can interact with more than one atomic nucleus simultaneously. The bonding
electrons have wave functions that resemble the atomic orbitals
near the nuclei but which build up in the region between the two
nuclei. The new resulting wave function is called a *molecular
orbital*. For the hydrogen molecular ion such a molecular orbital
has the simple configuration shown in Figure 3-3. The addition of a
second electron to this system will increase the binding of the two
nuclei but will not double it, since there will be some repulsion due
to the like charges of the electrons.

Two kinds of bonding molecular orbitals may be involved in more
complicated molecules: localized bond orbitals that contain only the
coordinates of the two nuclei (called $\sigma$ *orbitals*) and nonlocalized
orbitals involving two or more nuclei and perhaps even extending
over an entire large molecule (called $\pi$ *orbitals*). The larger the
nonlocalized orbital (i.e., the more spread out the electron probability distribution) the longer will be the wavelength for an electron
in that orbital. Of course longer wavelength means lower energy
($E = h\nu$) and lower energy implies more stability. In addition, lower
energy results from the fact that electrons are further apart in nonlocalized orbitals (i.e., less electron repulsion). Such large nonlocalized orbitals are found in conjugated ring structures like the
purines and pyrimidines, and they account for the stability of these
structures. Benzene is the best understood complex molecule in
terms of molecular orbital theory (Figure 3-4). The molecular $\pi$
orbitals can be described in terms of a fusion of atomic $p$ orbitals.
In like manner, the localized molecular bond orbitals, the $\sigma$ orbitals,
are formed from the overlap or fusion of atomic $s$ orbitals.

Two hydrogen $1s$ orbitals     Addition of two hydrogen
                               $1s$ orbitals to form $H_2^+$

FIGURE 3-3. Molecular orbitals for the hydrogen molecular ion.

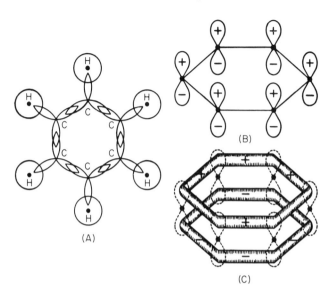

FIGURE 3-4.   The orbitals of benzene: (A) $s$ atomic orbitals; (B) $p$ atomic orbitals; (C) $\pi$ molecular orbitals, a fusion of the $p$ atomic orbitals. [From C. A. Coulson, *Quart. Rev. (London)* 1, 144 (1947).]

A single bond is usually a $\sigma$ bond and the molecular orbital is symmetrical about the axis of the bond. A double bond, on the other hand, may involve both a $\sigma$ and a $\pi$ orbital and will contain no axis of symmetry ($\pi$ orbitals change sign in passing through the plane of $\sigma$ bonds, as illustrated for benzene in Figure 3-4C). Although the strength of a $\sigma$ bond will not be affected by rotation about the axis of the bond, a $\pi$ bond will be seriously impaired by attempted rotation since the $p$ orbitals will be fused effectively only when the two halves of the molecule are properly oriented. Thus, there can be no free rotation about a double bond.

There are two basic approaches to the problem of chemical bonding. The Heitler-London, or valence bond approach, begins with the atoms and their intact atomic orbitals. The atoms are then moved closer and closer together and the properties of the resultant molecule are determined in terms of the perturbation of the atomic orbitals and the realization that two electrons are generally required to form a bond. The more useful and generally applied method is the Molecular Orbital (MO) approach which begins with the nuclear framework. The nuclei of the atoms are placed at their equilibrium separations and the electrons are then fed into the resulting force field, observing the Pauli exclusion principle. It is then possible by various approximation methods to calculate the molecular energy levels. It should be mentioned that while distinct in approach, both the valence bond and MO method yield the same result.

As a simple example of the MO approach, consider the orbitals possible in the formaldehyde molecule. In addition to the $\pi$ and $\sigma$ orbitals that we have discussed above, we must now introduce a new type of molecular orbital. It is the $n$ orbital and it closely resembles an atomic orbital in that an electron in this orbital is closely confined to the neighborhood of one nucleus and it interacts very little with other nuclei. Consequently, the $n$-orbital is a nonbonding orbital and it contains electrons that take small part in actual binding of nuclei. In formaldehyde an $n$-orbital is localized at the oxygen atom. Such orbitals are also found in heterocyclic compounds and in such they are always localized at the hetero atom. Thus, in pyrimidines the $n$-orbitals are located at one of the nitrogen atoms. Because of their loose coupling with the rest of the molecule the $n$-orbitals have energies that are higher than the delocalized $\pi$-bonding orbitals but lower than the orbitals for excited states. An energy level diagram for formaldehyde is shown in Figure 3-5. There are six bonding or valence electrons to be considered in this system and following the MO approach in principle we can place these six electrons in the six lowest possible energy states. (Note that it is only the valence electrons that need concern us in this process.) The situation in which the three lowest states are occupied by pairs of electrons is the ground state. The lowest lying excited states involve the reversal of one of the atomic $2p$ orbitals to generate a new set of *antibonding* $\sigma$ or $\pi$ orbitals (called $\sigma^*$ and $\pi^*$ respectively) in which there is a nodal plane rather than a buildup of electron density between the involved nuclei. In a molecule such as formaldehyde (since the interaction between $2p$ states is strongest in the axial direction) the ground $\sigma$ state and the excited $\sigma^*$ state will have the lowest and highest energies, respectively, in the energy level diagram. The presence of a nodal plane between bonded atoms will tend to weaken the bond, but that doesn't necessarily mean that the bond will break. This is especially true in a molecule like a pyrimidine in which there are six bonding $\pi$ electrons. The energy level

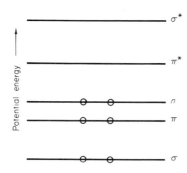

FIGURE 3-5.   Energy level diagram for formaldehyde. The symbols are defined in the text.

diagram as shown in Figure 3-5 can be useful for illustrating the transitions of electrons from one molecular orbital to another. However, it should be pointed out that such a diagram is not strictly correct since the actual energies of particular orbitals will depend upon which other orbitals have been vacated. Thus, the energy of an electron in the $\pi^*$ orbital will depend upon whether there is a vacancy in the $n$ or $\pi$ orbitals.

## 3-3.   Photon Effects on Orbital Electrons

The general effect of an incoming photon of the appropriate energy will be to promote an electron to an orbital of greater energy (e.g., from a bonding or nonbonding orbital to an antibonding orbital). The energies involved in these transitions between low lying molecular states are those of the visible and ultraviolet quanta, hence the importance of these spectral regions to photochemistry.

A $\pi\pi^*$ transition involves the excitation of a $\pi$ electron into a $\pi^*$ state. Likewise, an $n\pi^*$ transition involves the excitation of an $n$ electron into a $\pi^*$ orbital. $\sigma\sigma^*$ transitions can generally be ignored as a first approximation and need not concern us in the present discussion. Sigma electrons can not be neglected in detailed molecular orbital calculations, however, The $\pi\pi^*$ transitions are responsible for the most intense absorption bands in molecular spectra. This is due to a high degree of overlap between the ground state and excited state wave functions and to a strong dipole oscillation. Although the $\pi$ and $\pi^*$ orbitals differ in symmetry they cover approximately the same regions of the molecule. In contrast there is very poor spatial overlap between the localized nonbonding and the delocalized $\pi^*$ orbitals so that the transition probability for an $n\pi^*$ transition is correspondingly low. Because of this the absorption bands of $n\pi^*$ transitions are much less intense than those of $\pi\pi^*$ transitions by a factor of 10- to 100-fold. Also, the radiative lifetimes of $n\pi^*$ states should be greater than those of $\pi\pi^*$ states. However, the actual lifetime of the $n\pi^*$ state may be shortened considerably because of another property of these transitions: a large electron displacement that leads to a high degree of polarization of charge in the transition. Such a polarization leads to increased chemical reactivity (as an electron donor or acceptor) and makes the molecule especially susceptible to the deexcitation processes that involve chemical reaction.

## 3-4.   DISSIPATION OF PHOTON ENERGY

Of course most photochemical excitations of molecules do not lead to chemical reaction. There are a number of different pathways that the energy of absorbed photons may take after it has appeared as internal energy of molecular excitation. That the energy does not remain long in the absorbing molecule is illustrated by the fact that the color of most substances does not change during illumination. This must mean that the excited molecules return very quickly to the ground state where they can again absorb the same wavelengths as at the first instant of illumination.

What happens between the time a photon is absorbed and the return of the excited molecule to the stable ground state? Absorption occurs in about $10^{-15}$ seconds and the intrinsic lifetimes of $\pi\pi^*$ and $n\pi^*$ excited states are about $10^{-8}$ and $10^{-6}$ seconds respectively. The simplest way in which the energy may be dissipated is in the reemission of light. If this happens within $10^{-6}$ seconds after absorption it is called *fluorescence*. However, not all of the energy comes back as light and the fluorescent quantum is thus at a longer wavelength than the one that was absorbed. The energy deficiency can be understood in terms of the *Franck-Condon principle* which recognizes that the time required for the absorption of photons is about $\frac{1}{100}$ that of the period of vibration of a molecule. So nuclei do not appreciably alter their relative positions or kinetic energies during the act of absorption. The total energy of a molecule involves rotational and vibrational modes as well as electronic ones. The energies of the rotational modes are sufficiently smaller than the electronic ones so that they can be neglected, but those of vibration can not. To take a simple example, consider the ground state of a diatomic molecule. The overall energy can be plotted as a function of some molecular coordinate such as the internuclear separation. Different quantized modes of vibration can be represented as levels within such a potential energy curve as shown in Figure 3-6. The first excited electronic state (*singlet state*) can be represented by another potential energy curve on the same energy scale. It should be noted that, in general, the equilibrium separation of nuclei will be different in the excited state because of the changed electronic distribution which results in an altered charge distribution in the molecule. (In fact, any photon induced transition will result in some change in the charge distribution in the excited molecule.) As a consequence of the Franck-Condon principle the most probable electronic transitions are those for which the average interatomic distances are the same in initial and final states. Thus, the absorption of light does not necessarily bring the molecule to the lowest vibrational state in the new electronic configuration because the interatomic distances in this state may not be the same as in the ground state. Instead, the primary excitation will leave the molecule in a vibrating state in which interatomic distances correspond most closely to those in the ground state. The excess vibrational energy in the excited state will be dissipated as heat by impacts with the solvent and other molecules. Fluorescence will then generally occur from the lowest vibrational energy level of the excited singlet state as illustrated in Figure 3-6. Incidentally, this explains why the wavelength of fluorescence is independent of the wavelength of the exciting light. It is also evident from the foregoing that vibrational energy modes contribute to the spectral line width in absorption spectra.

   Some molecules, particularly those with highly conjugated structures have the ability to radiate from an electronic state that is intermediate between the ground state and the fluorescent state. This type of luminescence is called *phosphorescence* and for a number of cases this has been shown to be a triplet state. The *triplet state* is

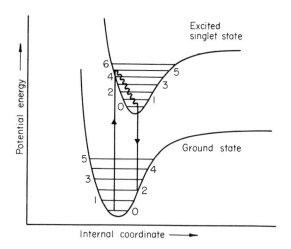

FIGURE 3-6. A diagramatic representation of photon absorption and fluorescence. A representative internal coordinate for a diatomic molecule might be internuclear distance. A potential well might then schematically represent the possible internuclear separations for a given electronic state of the molecule as shown in this figure. The minimum energy value or bottom of the well indicates the equilibrium internuclear separation. Numbered bands within these potential wells represent the different vibrational states of the molecule. (Rotational energy states could have been indicated in this figure as a fine structure of bands between the various vibrational levels.) At higher vibrational energy states the variation in internuclear separation increases until sufficient energy is attained to overcome the attraction between the nuclei, and this results in the breaking apart of the molecule. Note that the minimum internuclear distance approaches a limiting value as the vibrational energy increases. Vibrational deexcitation in the excited electronic state is indicated by the wavy line.

one in which the spin of the electron has flipped during the transition so that the system has two electrons with unpaired spins. No violation of the Pauli principle is involved, however, since the spatial quantum numbers for the two electrons are different. Transitions from the triplet state to the ground state are forbidden (i.e., flipping of the electron's spin upon returning to ground state is forbidden). As with many forbidden things this does not really mean that such a transition will never occur but merely that the probability of its occurrence is very small, such that the lifetime of a triplet state may be as long as $10^{-3}$ seconds or even a second. In terms of chemical reaction a second is a rather long time (in fact $10^{-3}$ seconds is a long time!). An important aspect of the triplet state is that it allows more time for chemistry to occur before the return to the ground state. Although there does not seem to be good theoretical justification for the temperature dependence of triplet state lifetimes, it turns out that phosphorescence is strongly temperature dependent, such that at very low temperatures lifetimes of the order of several hours can be observed.

As with fluorescence we can illustrate the phenomenon of phosphorescence in terms of an energy potential diagram as shown in Figure 3-7. Because the excited electron and its (former) partner have parallel spins, triplet states are of lower energy than their singlet counterparts. (Lower energy in this case results from a lower electron-electron repulsion by virtue of the greater spatial separation of the electrons in the triplet state.) Since the spin moment of an excited electron interacts with magnetic forces other than that of its ground state partner the probability of single-triplet transitions and consequently the lifetimes of the triplet states are markedly influenced by the electrical and magnetic environment of the excited molecules.

The transition from the singlet to the triplet state can be seen in the potential energy diagram (Figure 3-7) in terms of a point at which the potential curve for the singlet state crosses that for the triplet state. Collisional deactivation can drop the energy to this point and

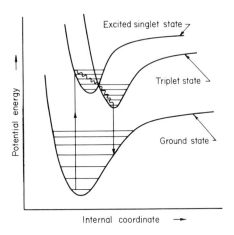

FIGURE 3-7. Potential energy diagramatic representation for triplet state excitation and phosphorescence.

if further collisions do not carry the molecule too rapidly past this point, transition to the triplet state may occur. Such nonradiative transitions between states of different multiplicity are called *inter-system crossing*. To restate: the obvious importance of the triplet state to photochemistry is its long lifetime. The longer a molecule is in an excited reactive state, the more likely that chemical reaction will occur.

There is another long-lived state that can sometimes occur following photon absorption and which leads to a delayed emission of a fluorescent energy photon. Such a state is not necessarily a triplet state, so it might be more generally referred to simply as a *meta-stable state*. Transitions from this state to the ground state are not merely forbidden but are apparently impossible. Therefore, the system must remain in this state until a quantum of sufficient energy is absorbed to restore it to the singlet state. Then the molecule can return to ground state by photon emission. Finally there is still another mode of delayed fluorescence observed in some organic crystals in which a population of molecules in triplet states can interact in such a way that two triplet molecules are converted into one molecule in the ground state and the other in the excited singlet state. The excited singlet then emits a photon in returning to the ground state.

We have discussed absorption and emission of photons in terms of oscillators. A useful parameter in the discussion of electronic absorption bands is that of the oscillator strength or absorption probability $(f)$. Classically, the oscillator strength measures the effective number of electrons whose oscillations give rise to a particular absorption or emission band. This dimensionless quantity is determined simply by integrating the area under an experimentally obtained absorption band. This area is then proportional to the absorption probability and related to it by a constant. Thus:

$$f = 4.32 \times 10^{-9} \int \epsilon(\bar{\nu}) \, d\bar{\nu}$$

where $f$ is the oscillator strength
$\epsilon$ is the molar absorptivity
$\bar{\nu}$ is the average wave number

The shapes of most absorption bands are such that the area can be approximated by the half-width of the band multiplied by the molar absorptivity.

The intrinsic probabilities of absorption and emission are proportional and thus, the intrinsic lifetime of an excited state varies inversely as the probability of absorption. The lifetime ($\tau$) is given by the relationship $\tau = k/\bar{\nu}^2 f$ where the constant $k$ is about 1.5 and $\bar{\nu}$ is the average wave number for the absorption band. Values of the oscillator strength range between 1.0 (very strong absorption bands) to less than 0.1 for very weak and sometimes forbidden transitions.

It is important to realize that the intrinsic lifetime is that which would prevail if the only means for leaving the excited state were fluorescence. The actual lifetime of the excited state will be less than the intrinsic lifetime because pathways other than fluorescence are avilable for vacating the state. We have mentioned phosphorescence as one such pathway.

It is also possible, however, to dissipate the excitation energy without emission of light at all. This process is called *internal conversion* or *vibrational deexcitation* and it involves the conversion of the energy of an excited electronic state to vibrational energy in a lower electronic state (Figure 3-8). The lower lying potential curve need not be a triplet state although it could be. There must be a crossing-over of two potential energy curves to allow internal conversion. The chemical consequences of vibrational deexcitation are

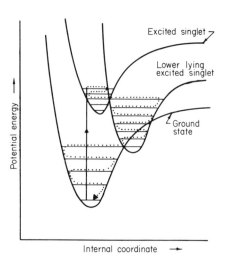

FIGURE 3-8. The nonradiative process of vibrational deexcitation and internal conversion.

important because the result is a "hot" molecule of much the same type that might be produced by a high temperature collision activation. Thus, photochemistry may result in the same products obtained in an ordinary chemical reaction, but with the photon replacing thermal agitation to supply the necessary energy of activation. One might wonder, then, why an ultraviolet photon of 4.9 ev (2537 Å) does not break chemical bonds if it is totally converted to vibrational energy, since it certainly carries more than enough energy to break most bonds. The evident reason is that the energy is not usually localized in a particular bond, but rather that it is distributed over many of the possible vibrational modes of many atoms in the molecule. Cases have been reported, however, where the absorbed electronic energy is transferred without dispersal to the vibrational energy of a particular bond, thus breaking the bond.

## 3-5. ROLE OF MOLECULAR ORIENTATION

At this point we should move from theory to experiment. There is a UV-induced photoproduct of thymine that consists of two thymine residues linked covalently through their respective 5 and 6 carbon atoms to form a dimer. This photoproduct was first demonstrated in irradiated frozen solutions of thymine by Beukers and Berends. Thymine in solution at room temperature is quite resistant to photochemical alteration. The reactivity of the thymine in frozen solution is explained in large part by the fact that the thymine residues are favorably oriented for dimerization. The importance of orientation may relate to the lifetime of the excited state. In solution the reaction would be diffusion controlled and the lifetime of the excited state of thymine under these conditions appears to be too short to allow for a high probability that two molecules of thymine will become properly oriented for dimer formation to occur. In support of this concept is the fact that thymine dimers formed in ice can be split to monomeric thymine by continued irradiation if the solution is allowed to thaw (see Section 4-5). The explanation is that as the dimers are split by UV (the formation and splitting of dimers are in equilibrium) the thymine monomers diffuse apart and are therefore not available for the reformation of dimers.

The yield of thymine dimers in solution can be enhanced if oxygen is excluded from the system. Oxygen and other paramagnetic substances quench the triplet state. In the absence of the quenching effect of oxygen, the lifetime of the triplet state is greatly lengthened, thereby giving the excited thymine molecules greater time to collide with other thymine molecules to form dimers. These results support the idea that the formation of the dimer can proceed through the triplet state of thymine. However, there is no oxygen effect on the production of thymine dimers in irradiated polynucleotides. In polynucleotides the thymine residues are already so close to one another that even in the presence of oxygen the dimerization rate is maximal. Under situations where the thymine residues are optimally oriented (i.e., thymine dimers that have been split in an ice matrix so that the thymine residues cannot diffuse apart) Lamola has shown that dimerization appears to go via the excited singlet state. Incidentally, oxygen does not quench the singlet state.

Although the orientation of molecules is important in the chemistry that follows photon absorption, it is also important in the photon absorption event itself. There is a net displacement of charge involved in any probable photon induced transition. This, of course, implies a direction for the excitation and in fact, an induced dipole. The direction and magnitude of this induced dipole is expressed in the vector quantity, the *transition moment*. The preferred direction for excitation is a direct consequence of the fact that the dipole has a preferred direction for absorption. The direction of the transition moment can sometimes be predicted from theory, but it is very difficult to determine it experimentally since determinations must be made on well-oriented systems (i.e., crystals) using polarized light. There will be preferred directions for absorption depending upon the particular transitions allowed. Figure 3-9 shows the calculated transition moments for the four bases in DNA. Note that there are at least two $\pi\pi^*$ transitions possible in each base. (The $n\pi^*$ transitions are perpendicular to the plane of the base and are much weaker.) The $\pi\pi^*$ transition is considerably stronger in heterocyclics ($f = 0.20$ in thymidine) than in benzene ($f = 0.0014$).

An elegant determination of the transition moments of purines and pyrimidines has been performed upon single crystals of 1-methyl thymine and upon a mixed crystal of 1-methyl thymine and 9-methyl adenine by Stewart and Davidson. Thin sections ($0.1\mu$ thickness) were cut from these crystals by ultramicrotome and then examined in polarized ultraviolet light. Two $\pi\pi^*$ transitions were determined, at 14° and 19° respectively, from nitrogen atom one in

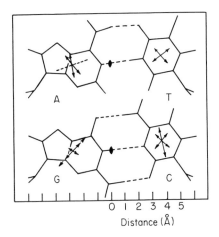

Distance (Å)

FIGURE 3-9.  Transition moments ($\pi \to \pi^*$) of the DNA bases as determined from molecular orbital calculations. Relative strengths are indicated by the lengths of the arrows shown in the planes of the bases. [Redrawn from H. DeVoe and I. Tinoco, *J. Mol. Biol.* 4, 518 (1962).]

thymine and these positions were in reasonable agreement with published calculations (Figure 3-9).

## 3-6.  HYPOCHROMICITY

As discussed in Section 1-3, any perturbation of the electric field environment of a molecule would be expected to alter the intensity of absorption bands. One would expect an effect upon absorption, for example, if one were to orient absorbers so that their transition moments were aligned. The mechanical analogy of this effect is that of coupled pendulums, or coupled oscillators in general. The setting of one oscillator into motion will induce a dipole in that direction and will, in turn, set adjacent coupled oscillators into motion. For such coupling between oscillators, the effect drops off as the reciprocal of the distance cubed, so that if this discussion were extended to the case of oriented bases in DNA (the planes of the bases are perpendicular to the long axis of the molecule) then 5 base pairs should exhibit about 60% of the hypochromism of an infinite helix. One turn of the helix (10 base pairs) should exhibit roughly 80% of the hypochromism of an infinite helix. Thus, only a relatively small number of bases in a polynucleotide strand are required to approximate the effect in a large polymer. The fact that hyperchromism is also observed upon the hydrolysis of single stranded DNA indicates that some base interaction (i.e., stacking) occurs in the single-strand configuration (Figure 1-8).

Although it is possible to calculate an expected hypochromism on the basis of coulombic interactions of the stacked bases alone, it has also been demonstrated that solvent interactions with the absorbers can give rise to hypochromism. For example, inosine exhibits less absorbance in the nonpolar solvent, acetonitrile, than in water. When the absorbance of isotactic polystyrene is compared to that of atactic or disordered configurations of this polymer in a nonpolar solvent, some agreement is seen with the theory that coulombic interactions of monomers is responsible for hypochromism. However, an additional hypochromic effect is seen if the comparison is carried out in a polar solvent.

It is evident that at least two factors are involved in the phenomenon of hypochromicity in DNA: (1) The partial alignment of transition moments by base stacking, and (2) the additional enhancement of the coupling of such interactions in polar solvents. The solvent itself may also affect the base stacking since this is a *hydrophobic* interaction. Such stacking is evident even for dinucleotides in polar solvents, but is not as evident if they are in nonpolar solvents. The synthetic polynucleotide, polyribouridylic acid (poly rU), does not stack at all in aqueous solution above about 16°C; consequently it exhibits very little hyperchromicity. On the other hand polyriboadenylic acid (poly rA) does stack to a significant degree and correspondingly exhibits a rather large hyperchromic effect when the stacking is disrupted thermally. Since poly rA also forms a multistranded hydrogen-bonded complex with itself, one might wonder what the contribution of hydrogen bonding might be to the hyperchromic effect. Evidence that this contribution is negligible is seen in the observation of hyperchromicity with polyribohydroxyethyladenylic acid which is sterically unable to form hydrogen-bonded base pairs. Thus, in DNA it is likely that the double-helix configuration is important to hypochromicity only by stabilizing the orientation of the stacked bases in the individual polynucleotide strands.

Although hypochromism cannot be used in a simple way to look at the compositional character of DNA it is possible to demonstrate the selective "melting" out of regions high in adenine-thymine content or guanine-cytosine content by studies on the wavelength dependence of thermal denaturation curves (Figure 3-10).

Since an absorption spectrum includes the sum total of all of the different allowed transitions, there is the interesting possibility that one might observe a hyperchromic effect and a hypochromic effect at the same time for the mutually perpendicular $n\pi^*$ and $\pi\pi^*$ transitions in the bases. This may be possible for the case of the two-stranded complex of poly rA and poly rU. Upon formation of this complex there is a marked hypochromicity at 260 nm ($\pi\pi^*$ transition), but a hyperchromicity at 280 nm which has been postulated to be due to the $n\pi^*$ transition. On the other hand it is not absolutely certain that the 280 nm band is due to the $n\pi^*$ transition, especially since poly dAT (an alternating copolymer of deoxythymidine and deoxyadenine residues) does not show this effect and in fact gives the same ratio of hypochromicity over the entire spectrum.

Hypochromism can also be demonstrated in proteins. Thus, poly-L-lysine exhibits a hypochromic effect at 195 nm upon going from the random coil form at pH 6 to the alpha helix at pH 10.8. If heated to 52°C, however, the alpha-helical configuration is disrupted and the semiplanar beta form occurs with a pronounced hyper-

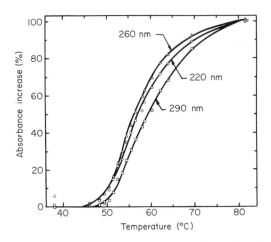

FIGURE 3-10. Absorbance increase at three wavelengths as a function of temperature in calf thymus DNA at pH 6.5. [Adapted from G. Felsenfeld and G. Sandeen, *J. Mol. Biol.* **5**, 587 (1962).]

chromicity. The hyperchromic effect might be used in the determination of the helical content of proteins were it not for the interfering absorption bands of the aromatic amino acids.

## 3-7. ENERGY TRANSFER

There is another consequence of the coupling of absorbers that is of importance to photochemistry and that is the possibility of energy transfer. This represents a new mode for dissipation of absorbed photon energy which might be called "passing the buck." A photon is absorbed in molecule A, the excitation energy is then transferred to molecule B, and molecule B must then dispose of it. All of the usual avenues for dissipation of this energy are then available, as discussed in Section 3-4. If the energy of molecule B is dissipated by fluorescence, the process is called *sensitized fluorescence*. It is best observed in situations in which the fluorescent spectra and absorption spectra of molecular species A and B are different. Such an energy transfer process is distinct from the trivial possibility that a fluorescent quantum emitted by molecule A might be absorbed and then reemitted by molecule B. The transfer process is in competition with the other pathways for deexcitation. The time required for such transfer can be much shorter than fluorescence lifetimes and the efficiency of transfer can be much greater than even the maximum efficiency of fluorescence of the primary absorber. Sensitized fluorescence has been demonstrated in a number of dye complexes and the distance of energy transfer has been shown to sometimes exceed 100 Å although it never exceeds the wavelength of the sensitizing photons. *Sensitized phosphorescence* has also been demonstrated. It is not really necessary to think of the energy transfer process as involving the localization of the incident energy in one unit and then the subsequent transfer of this packet of energy to another unit. If there are strong electric field interactions between the units, then the transfer can be described in terms of a delocalized excitation energy that is distributed over the entire ensemble of units.

Evidence for energy transfer in DNA is the fact that the phosphorescence of native DNA is not the sum of the expected emission from individual nucleotides. Using electron spin resonance and optical emission it has been demonstrated that the emission is predominantly from thymine. Samples were dissolved in an ethylene glycol-water glass at 77°K in order to stabilize triplet and other metastable states. It was found that at neutral pH, adenylic acid (AMP) and guanylic acid (GMP), but not thymidylic acid (TMP) or cytidylic acid (CMP), exhibit phosphorescence. The decay times were of the order of 1 and 2 seconds, respectively, for GMP and AMP. However, the observed decay time in DNA phosphorescence was only 0.3 seconds. Also, the quantum yield for phosphorescence in DNA was only 0.002 as compared with 0.02 for GMP. Poly dAT gave phosphorescence spectra and decay times quite similar to those for DNA although it was apparent that adenine contributed less than 15% of the observed phosphorescence. This all pointed very strongly to thymine as the phosphorescing moiety. Upon closer examination of the photochemistry of thymine it was found that a triplet state with a 0.5 second lifetime could be obtained when the irradiation was performed at pH 9.8. Under these circumstances a proton is removed from nitrogen-1 position of thymine. Since this proton is involved in the normal hydrogen bonding of thymine to adenine in DNA, it might be expected that this proton could be shifted to adenine with the expected thymine excitation in DNA. In summary, it is apparent that photons absorbed in either thymine or adenine in DNA can be equally effective in promoting thymine phosphorescence.

To anticipate our later discussion of biological effects of photons, we might cite the interesting example in which energy transfer apparently provides a protective effect in a biological inactivation process. The presence of the fluorescent dye, acridine orange, during UV irradiation enhanced the survival of *E. coli* strain B/r. It apparently acted as an energy sink for the harmless dissipation by sensitized fluorescence of photons absorbed in DNA. The enhanced survival of the bacteria was found to be somewhat greater if the irradiation was carried out in a nitrogen atmosphere rather than in the presence of oxygen. Perhaps an additional inactivation effect is occurring in the presence of oxygen by photodynamic action (Chapter 9).

It should be emphasized again that environmental conditions which alter the lifetimes of the excited states change the distribution of decay modes and the resultant photochemistry. It is conceivable that the quenching of triplet state excitation could reduce

UV mutagenesis in a biological system by reducing the yield of a particular photoproduct. In the next chapter we will give a number of specific examples of the effects of environment upon photochemical reactions.

## GENERAL REFERENCES

D. Cram and G. Hammond, "Organic Chemistry," Chapters 5 and 27. McGraw-Hill, New York, 1959.

G. M. Barrow, "Introduction to Molecular Spectroscopy." McGraw-Hill, New York, 1962.

J. B. Thomas, "Primary Photoprocesses in Biology." Wiley, New York, 1965.

R. K. Clayton, "Molecular Physics in Photosynthesis," Chapter 7. Ginn (Blaisdell), Boston, Massachusetts, 1965.

J. C. Speakman, "Molecules." McGraw-Hill, New York, 1966.

J. G. Calvert and J. N. Pitts, Jr., "Photochemistry." Wiley, New York, 1966.

N. J. Turro, "Molecular Photochemistry." Benjamin, New York, 1967.

# 4

# Photochemistry of the Nucleic Acids

## 4-1.  INTRODUCTION

Many of the biological effects of UV irradiation can now be explained in terms of specific chemical and physical changes in DNA. Correlations have been made between the survival of UV-irradiated cells and the production of certain types of photochemical damage in their DNA. It is also becoming increasingly apparent that DNA does not exhibit the same sensitivity to UV under all experimental conditions. The intrinsic sensitivity of DNA to photochemical alteration can be changed by a variety of biological (e.g., growth state

57

of cells), chemical (e.g., base analog substitution), and physical (e.g., denaturation; freezing; drying) techniques. To give an example: one photoproduct that is produced in high yield and appears to be the major cause of death in irradiated bacterial cells is not produced to a significant extent in bacterial spores. Different types of photoproducts must inactivate irradiated vegetative cells and spores, respectively. Simple generalizations, therefore, cannot be made as to which photoproduct in DNA is the most important to all irradiated cells under all experimental situations.

In addition to the intrinsic sensitivity of the DNA, we must also consider the ability of the cell to repair the damage. The repair of radiation damage is the subject of Chapter 7 but will also be mentioned here where necessary for clarity. The point to be kept in mind, however, is that if a lesion is completely repaired by a cell it cannot be of biological importance to that cell, while lesions that are not repairable (or are not repaired) may be of biological importance. In this chapter we will summarize the different photochemical lesions that are currently known to be produced in nucleic acids and to assess where possible their biological importance.

4-2.  EFFECTS OF UV ON RIBOSE AND DEOXYRIBOSE

Although carbohydrates make up about 41% by weight of the nucleic acids, they show essentially no UV absorption at wavelengths above 2300 Å, and therefore would not be expected to undergo photochemical reactions when irradiated with light of wavelengths greater than 2300 Å. The photochemistry of carbohydrates has not been systematically studied in recent years. Earlier reports on the photochemical alteration of carbohydrates are questionable because of the failure to adequately filter out wavelengths of light below 2300 Å.

An interesting indirect effect of UV on deoxyribose has been reported. A cell is much more sensitive to killing by UV if 5-bromouracil (5BU) replaces the thymine in its DNA (for further discussion of the effects of analog substitution see Section 4-11). One effect of UV radiation on 5BU in DNA is debromination with the consequent production of a uracil radical. In the absence of an added hydrogen donor (such as cysteamine) a hydrogen atom is abstracted from the adjacent deoxyribose. This leads to the production of uracil, to the destruction of the deoxyribose (by mechanisms presently unknown) and ultimately to a chain break in the DNA. Thus, under certain conditions, chemical alterations in the carbohydrates of the nucleic acids can be brought about when the nucleic acids are exposed to UV even though the primary absorption of photons does not occur in the carbohydrates.

## 4-3. EFFECTS OF UV ON PURINES

Purines are approximately ten-fold more resistant to photochemical alteration than are the pyrimidines ($\Phi$ for purines $\sim 10^{-4}$; pyrimidines $\sim 10^{-3}$). Because of this difference in sensitivity to photochemical alteration, it has been implied that the photochemistry of the purines is not important biologically, since by the time a significant amount of purine damage has occurred, the cells would have been inactivated by pyrimidine damage. Although statistically this hypothesis has much in its favor, the biological importance of purine photochemistry should not be so quickly dismissed. Although the absorption of UV by the purines does not result in the photochemical alteration of the purine ring with a high efficiency, some of the absorbed energy may well be transferred to the pyrimidines or to the sugar-phosphate backbone of DNA and thus result in chemistry. The transfer of energy from adenine has been implicated in the formation of a thymine radical in UV-irradiated poly-dAT (see Section 3-7).

With the current availability of radioactive-labeled purines and of purine oligonucleotides, the reinvestigation of the photochemistry of the purines would seem to be a fruitful venture. In view of the light intensities required for the study of the photochemistry of the purines, the technique of flash photolysis should be a most useful tool.

## 4-4. HYDRATION PRODUCTS OF THE PYRIMIDINES

When solutions of uracil and cytosine and their derivatives are irradiated, they lose their characteristic UV absorbance but this can be largely regenerated by heat, alkali, or acid treatment (Figures 4-1 and 4-2). It was postulated that the hydration of the 5–6 double bond of uracil could account both for the loss of the ab-

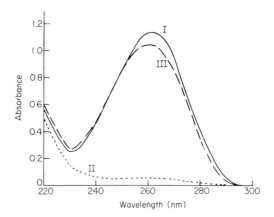

FIGURE 4-1.   Effect of UV irradiation and subsequent addition of acid on the spectrum of uridylic acid. I, Absorption spectrum before irradiation; II, absorption spectrum after 40 minute irradiation; and III, absorption spectrum after reversal at pH 0.8 (42 hours at pH 0.8 at 20°C). [From R. L. Sinsheimer, *Radiation Res.* 1, 505 (1954).]

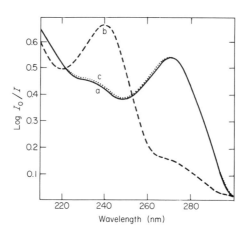

FIGURE 4-2.   Reversible photolysis of cytidine or cytidylic acid at neutral pH: (a) before UV irradiation, (b) after irradiation, and (c) irradiated solution after 16 hours at room temperature or 5 min at 80°C. [From D. Shugar, *in* "The Nucleic Acids" (E. Chargaff and J. N. Davidson, eds.), Vol. 3, p. 39. Academic Press, New York, 1960.]

sorption spectrum and the reversibility by subsequent treatment with acid or heat. The ultimate proof of this postulate came when 6-hydroxy, 5-hydrouracil (Figure 4-3) was synthesized and was shown to be identical with the reversible radiation product of uracil (note that 5-hydroxy, 6-hydrouracil is stable to heat). Direct chemical evidence for the photochemical hydration of cytosine derivatives is now also available. There is no direct evidence for the production of hydrates of thymine although several reports are suggestive of its transient formation. The formation of uracil hydrate has been shown to occur via the excited singlet state, as evidenced by the fact that oxygen does not quench the reaction and that no hydrates are formed when uracil is raised to the triplet state by molecular photosensitization (defined in Section 4-5).

The formation of the water addition photoproduct of cytosine in irradiated denatured DNA has been inferred from the appearance of a heat-reversible absorption peak around 2400 Å. Dihydrocytosine derivatives exhibit a characteristic absorption peak at this wavelength (see Figure 4-2). Irradiated native DNA, however, showed no such heat-reversible absorption peak. Evidence for reversible hydrate formation has been obtained in irradiated polyribocytidylic acid (poly-rC) but none has been found in the homocopolymer composed of polydeoxyinosine hydrogen-bonded to polydeoxycytidylic acid (poly-dI:dC) or in poly-rI:rC. These data suggest that hydrates of cytosine are probably not formed in irradiated double-stranded DNA but they are formed in single-stranded DNA. More sophisticated techniques are being developed (e.g., isotope exchange reactions) to quantitate the production of hydrates in polynucleotides and preliminary results indicate a small amount of hydrate formation in irradiated "native" DNA. One must be cautious in such studies, however, because other photoproducts (such as dimers) can distort the DNA and thus effectively denature it locally.

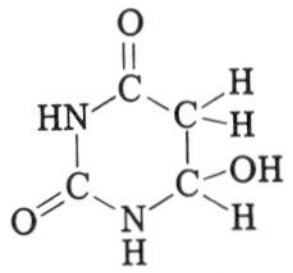

FIGURE 4-3.  6-Hydroxy, 5-hydrouracil.

During replication and/or transcription of the DNA, there may be short regions of single-strandedness, and in these regions the formation of pyrimidine hydrates may well be of importance. The possible role of pyrimidine hydrates in causing mutations has been demonstrated by Grossman in an *in vitro* model system. When polycytidylic acid was irradiated with UV, its coding properties in an RNA polymerase system were altered. The irradiated polymer lost its ability to code for the incorporation of guanylic acid, but it then coded for the incorporation of adenylic acid. The increase in adenylic acid incorporation was heat reversible under conditions known to reverse pyrimidine hydrates and for this reason it was suggested that the code change might be the result of the formation of cytosine hydrates. The formation of hydrates in single-stranded regions of the DNA may well be of significance in the production of mutations, which may or may not be lethal.

4-5. CYCLOBUTANE-TYPE DIMERS OF THYMINE, CYTOSINE, AND URACIL

When an aqueous solution of thymine is irradiated (2537 Å) it loses its characteristic absorption properties at a rate about $\frac{1}{10}$ that of uracil ($\Phi = 0.4 \times 10^{-3}$ for thymine; $5.5 \times 10^{-3}$ for uracil), and the photoproduct(s) formed cannot be reversed by acid or heat as is the case for uracil. If a solution of thymine is frozen and then irradiated it shows a greatly increased photochemical reactivity ($\Phi = \sim 0.2$) and the major product formed is a dimer. The probable effect of freezing is to bring the thymine molecules into an oriented juxtaposition favorable for a bimolecular photochemical interaction. Dry films of thymine also yield the same photoproducts obtained in frozen solution but the yield is much lower.

Although the dimer cannot be reversed by acid or heat it can be reversed to monomeric thymine by irradiation with light of short wavelength (see below).

To form the thymine dimer (Figure 4-4), two thymine molecules are linked to each other between their respective 5 and 6 carbon atoms thus forming a cyclobutane ring between the two thymines. There are six possible isomers of the thymine dimer and these have been isolated from irradiated thymine oligomers. The Type I (cis-syn) thymine dimer (Figure 4-4) is the one formed between adjacent

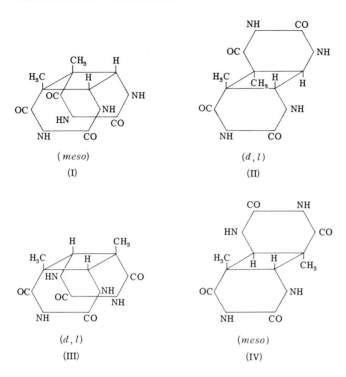

FIGURE 4-4. Isomeric forms of the cyclobutane-type thymine dimers. [Modified from D. L. Wulff and G. Fraenkel, *Biochim. Biophys. Acta* **51**, 332 (1961).]

thymines in the same strand of DNA. Certain of these isomers are stable to acid hydrolysis while others are not. This is an important point because acid hydrolysis is the usual method for liberating photoproducts from irradiated DNA. Acid hydrolysis may, therefore, preclude the observation of certain labile photoproducts. Milder methods of hydrolysis, perhaps enzymic, must be developed. This may be a difficult task since UV-irradiated DNA is rather refractory to enzymic digestion.

There is a wavelength dependency for the formation and monomerization of the cyclobutane-type thymine dimer (Figure 4-5), such that after a sufficient dose of UV a photosteady state between monomer and dimer is reached that is characteristic for the wavelength used (Figure 4-6). At the longer wavelengths (around 2800 Å) the formation of the dimer is favored while at the shorter wavelengths (around 2400 Å) monomer formation is favored. This response is due to differences in the absorption spectra of thymine and its dimer (Figure 4-7) and in the quantum yields for the formation and splitting of the dimer (Figure 4-5).

Five other dimers of the natural pyrimidines are also known. These are the dimers of uracil, cytosine, uracil–thymine, cytosine–thymine, and uracil–cytosine. Because these dimers also have the property of short wavelength reversal they are believed to have the same skeletal structure as the cyclobutane-type thymine dimer.

As with the thymine dimer the short wavelengths (around 2400 Å) are also more efficient in monomerizing the uracil dimers, although the uracil reaction is complicated by the production of hydrates which interfere with dimer production. The isolation of cytosine dimers is complicated not only by the competition of the hydrate reaction but also by the fact that cytosine deaminates readily when

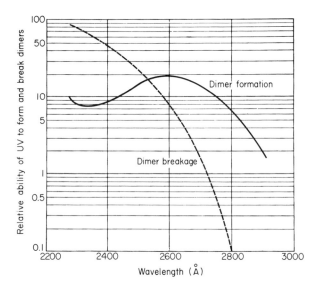

FIGURE 4-5. The efficiencies for forming and breaking cyclobutane-type thymine dimers in polythymidylic acid are both wavelength dependent. Those wavelengths longer than 2540 Å are more efficient in forming dimers; those shorter than 2540 Å are more efficient in breaking dimers. [From R. A. Deering, *Sci. Am.* **207**, 135 (1962).]

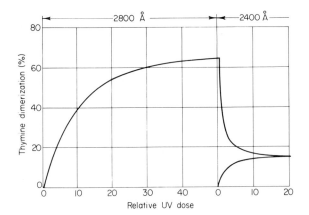

FIGURE 4-6.   The extent of thymine dimerization varies with wavelength and dose in polythymidylic acid. With increasing exposure at 2800 Å the proportion of thymine units that are dimerized increases until an equilibrium state is attained in which 65% of the thymine units are dimerized. Irradiation at 2400 Å of the same or of a different sample results in a new equilibrium level with about 17% of the thymines dimerized. [From R. A. Deering, *Sci. Am.* **207**, 135 (1962).]

its 5–6 double bond is saturated. Cytosine dimers are therefore readily converted to uracil dimers. Cytosine dimers are formed at lower rates than are thymine dimers but they are monomerized at more rapid rates by short wavelength radiation than are thymine or uracil dimers. It is apparent that if a cytosine dimer in the DNA of a cell were to deaminate to form a uracil dimer and if this dimer were then split *in situ* by the photoreactivating enzyme (Section 7-3), a mutation could result. The resultant uracil residues would base pair with adenine rather than with guanine.

Up to now we have talked about photochemical reactions that occur as the result of the direct absorption of photons by the reacting species. Thymine dimers can be formed, however, by wavelengths of light that are not absorbed by thymine providing that the thymine is in the presence of suitable molecules that do absorb these wavelengths. This process is called *molecular photosensitization*. It requires that the triplet state of the absorbing species (the photosensitizer) be slightly higher in energy than the triplet state of the thymine. Upon collision, the triplet energy of the photosensitizer is transferred to the thymine, yielding thymine in its triplet state with the subsequent possibility for the formation of thymine dimers. Examples of this situation are the formation of thymine dimers by light of wavelengths above 3000 Å when the thymine is dissolved in acetone, and the formation of thymine dimers in DNA when it is irradiated with wavelengths above 3000 Å in the presence of $10^{-2}M$ acetophenone. One advantage of the use of molecular photosensitization to drive a reaction is that it can be performed at wavelengths where the reverse reaction (dimer splitting) does not occur. With this technique one should be able to achieve essentially a quantitative conversion of adjacent thymine residues to dimers, rather than achieving an equilibrium yield as obtained by the direct excitation of thymine residues. It has long been known that cells can be both mutated and inactivated by visible light. It is presently not known how these effects are mediated. They could either be caused by photodynamic mechanisms (Chapter 9) or by molecular photosensitized reactions of the type just described.

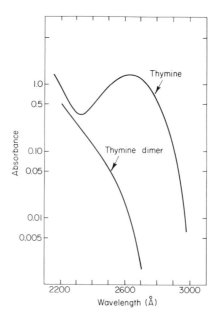

FIGURE 4-7. The absorption spectra of thymine (25 μg/ml) and of purified thymine dimer (34 μg/ml). [Modified from R. Setlow, *Biochim. Biophys. Acta* 49, 237 (1961).]

Certain of the lines of evidence that indicate that the thymine dimer is of biological importance are the following:

1. The short wavelength reversibility of the thymine dimer has been used to show that much of the inactivation of transforming DNA that is brought about by irridiation at 2800 Å is reversed by a second irradiation at 2400 Å. In this manner, it has been determined that after high doses of UV at 2800 Å about 50% of the inactivation of *Hemophilus influenzae* transforming DNA can be attributed to the production of thymine dimers.

2. Bacterial cells that have been irradiated with UV show an increased survival if they are additionally irradiated with visible light. This process is known as photoreactivation (Section 7-3). The enzyme responsible for this phenomenon has been isolated and shown to be specific for the repair of cyclobutane-type pyrimidine dimers. If this enzyme has the same specificity *in vivo* then it argues for the biological importance of pyrimidine dimers. It should be emphasized, however, that photoreactivation does not cause the complete reversal of UV damage to a cell, suggesting that there are biologically important photochemical lesions produced in DNA besides the cyclobutane-type pyrimidine dimers.

3. Certain strains of *E. coli* are very sensitive to killing by UV while others are very resistant. All the resistant strains tested have the ability to excise thymine dimers from their DNA and to undergo repair replication of their DNA whereas the sensitive strains tested were unable to perform this "cut and patch" type of repair (Section 7-4). While the pyrimidine dimers may be of importance to an organism that is incapable of repairing them, they are probably of less importance to those organisms that have efficient repair mechanisms.

Since much of the work to prove the biological importance of thymine dimers was done prior to the general recognition of the involvement of cytosine in cyclobutane-type dimers some of the biological effects assigned to thymine dimers by the above experiments should probably be re-assigned to dimers involving cytosine.

The majority of our knowledge on the photochemistry of the nucleic acids concerns the thymine dimer and the largest amount of evidence supporting the biological importance of a given photoproduct is also concerned with the thymine dimer. The sheer volume of these data have tended to imply that other types of photochemical lesions in DNA are not of biological importance. The thymine dimer is unquestionably of major biological importance under many experimental conditions but it is not of major importance in all situations (see Section 4-9). The photochemical yield and relative biological importance of the thymine dimer is different for different cells and can even change for a given cell under different growth and irradiation conditions. Therefore, it must be remembered that the biological importance of any photoproduct depends upon (1) whether or not it is formed under a particular set of experimental conditions and (2) if formed, whether or not the particular system under study is capable of repairing the lesion.

4-6. OTHER PHOTOCHEMICAL REACTIONS OF THE PYRIMIDINES

Many pyrimidine photoproducts other than the hydrates or the cyclobutane-type dimers are produced both *in vivo* and *in vitro*. Some of these appear to be monomeric while others appear to be dimeric in nature. The structure of these new photoproducts is largely unknown and their identity is based upon chromatographic properties (using radioactive labels). The relative biological importance of these photoproducts also remains to be determined.

The tentative structures of certain noncyclobutane-type photoproducts of thymine are given in Figure 4-8. Compound A is a minor product formed when thymine is irradiated in frozen solution and subsequently heated in acid. Compound B is a minor product formed when thymidylyl–thymidine (TpT) is irradiated in solution. Compound C has been isolated from irradiated DNA. Compound D has been isolated from irradiated bacterial spores and from DNA that has been irradiated in frozen solution.

The spore photoproduct deserves special consideration. After high doses of UV ($2 \times 10^5$ ergs/mm$^2$ at 2537 Å) about 30% of the thymine can be converted to this product. Although the spore photoproduct is a type of thymine dimer, it does not exhibit the short wavelength reversal properties of the cyclobutane-type thymine dimers. Therefore, the yield of this product can approach the maximum determined by the number of thymines that are nearest neighbors in the DNA. The maximum yield of cyclobutane-type dimers is additionally dependent upon the equilibrium between the

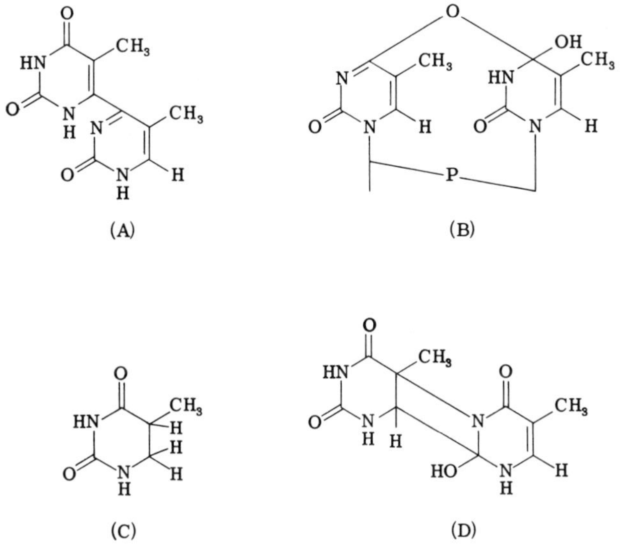

FIGURE 4-8.    Tentative structures of certain noncyclobutane-type photoproducts of thymine. (A), 6,4'-{5'-methylpyrimidin-2'-one}-thymine; a minor product formed when thymine is irradiated in frozen solution and subsequently heated in acid. (B), TpT⁴, a minor product formed when thymidylyl-thymidine (TpT) is irradiated in solution. (C), dihydrothymine, has been isolated from irradiated DNA. (D), azetane thymine dimer, has been isolated from irradiated bacterial spores and from DNA that has been irradiated in frozen solution.

formation and splitting of dimers (Section 4-5). Thus, in irradiated *E. coli*, where 30–40% of the thymines could theoretically dimerize only about 15% are dimerized after high doses of UV at 2537 Å.

During the germination of spores the yield of spore photoproduct decreases and the yield of cyclobutane-type pyrimidine dimers increases. The sensitivity to killing increases concomitantly. In fact, the cyclobutane-type pyrimidine dimers appear to be about eleven times more effective in killing vegetative cells than are the spore photoproducts in killing spores. This could imply that the spore photoproduct is more efficiently repaired than are the cyclo-butane-type dimers. The spore photoproduct does disappear from germinating spores but has not been recovered in acid extracts of the spores (as are the cyclobutane-type dimers from excising strains of vegetative cells). The molecular mechanism for this repair is unknown at present.

In the normal (wet) or *B* conformation of DNA the planes of the bases are parallel to each other and perpendicular to the helical axis. Under these conditions the formation of the cyclobutane-type dimer seems to be favored. In the dry, or *A* configuration (spores are considered dehydrated relative to vegetative cells) the planes of the bases are still parallel but are inclined 70° to the axis of the helix. This latter conformation of DNA would favor the formation of the azetane-type thymine dimer (Figure 4-8D).

The pyrimidines will react photochemically with other compounds. Alcohols (R–OH), cysteine (R–SH) (see Section 4-9) and hydrogen cyanide (H–CN) photochemically add across the 5–6 double bond of the pyrimidines. Since the addition products of methanol and ethanol can be reversed by acid they appear analogous to the water (H–OH) addition product described in Section 4-4 and therefore are expected to be attached to position 6. HCN and cysteine are known to add at the 5 position and these are stable products. The chemically synthesized isomer of the photohydrate of uracil [5-hydroxy, 6-hydrouracil] is also stable to acid.

4-7.  EFFECT OF UV ON THE MOLECULAR WEIGHT OF DNA
      (CHAIN BREAKAGE)

Ultraviolet radiation can reduce the molecular weight of DNA. However, no dramatic effect of DNA base composition upon the efficiency of DNA chain breakage brought about by UV irradiation *in vitro* has been observed. The photochemical lesion leading to chain breakage in natural DNA is unknown. We have previously discussed (Section 4-2) a mechanism whereby the substitution of thymine residues by 5BU can sensitize DNA to UV induced chain breakage. The dose of UV required to reduce the molecular weight of *Diplococcus pneumoniae* DNA by 50% was about 100 times that required to reduce the transforming activity of the streptomycin marker in this DNA to the same extent. At the dose of UV required to kill 99% of a population of phage T7, no chain breaks were detected. Therefore, current evidence suggests that at low doses of UV, chain breakage may occur too infrequently to be of biological importance.

Attempts to measure the number of DNA chain breaks produced when bacterial cells are irradiated with UV are complicated by the fact that UV resistant strains can cut out base damage (dimers, etc.) from their DNA thus indirectly producing a chain break. It would be necessary to use strains that are unable to perform this "cut and patch" type of repair if one wanted to measure accurately the number of chain breaks produced as an immediate consequence of the absorption of UV radiation. The method currently favored for the measurement of chain breaks in DNA is zone sedimentation in alkaline sucrose gradients in an ultracentrifuge.

## 4-8. DNA-DNA CROSS-LINKS

DNA-DNA cross-links leading to gel formation have been observed in DNA irradiated while dry and in UV-irradiated salmon sperm heads where the DNA is known to be very tightly packed, but they have *not* been observed in UV-irradiated wet cells. It is interesting therefore that although pyrimidine dimers appear to be involved in the formation of DNA-DNA cross-links (i.e., irradiated dry pyrimidine oligonucleotides form cross-links but purine oligonucleotides do not) the conditions that favor their formation are conditions that do not favor the formation of the cyclobutane-type thymine dimers. DNA-DNA cross-links appear to be of little biological importance to normal wet cells, but may achieve a position of greater biological importance when cells, viruses, or transforming DNA are irradiated dry.

Another type of DNA–DNA cross-linking causes the two strands of a molecule of DNA to be connected so that they can no longer be separated when the DNA is denatured with heat or formamide. For a given dose of UV the extent of cross-linking is greatest at the temperature where DNA is 20% denatured, and for a given temperature, it has been shown to increase proportionally with the adenine-thymine content of the DNA. The latter response suggests that some type of thymine dimer might be responsible for the cross-linking. The structure of DNA suggests that the cyclobutane-type thymine dimer that is most likely involved in these cross-links is structure IV in Figure 4-4. This dimer, however, is rapidly destroyed by the acid hydrolysis conditions used to isolate the more common cyclobutane-type thymine dimer. This lability to acid hydrolysis probably explains why the suspected cross-linking dimer has not yet been isolated from irradiated DNA. DNA in which almost all of the thymine was replaced by bromouracil was about five times more sensitive to interstrand cross-linking by UV than was normal DNA. Since no interchain cross-links were detected in normal phage T7 irradiated to a survival of 1% the biological importance of this lesion seems in doubt at low doses of UV. However, this lesion may achieve a position of greater biological importance at higher doses of UV in those cells that are relatively resistant to ultraviolet radiation.

4-9. THE CROSS-LINKING OF DNA TO PROTEIN

There is a progressive decrease in the amount of DNA that can be extracted from bacterial cells following increasing doses of UV. To demonstrate this effect the cells were lysed and the proteins were denatured by treatment with a detergent (sodium lauryl sulfate). The subsequent addition of 1 $M$ KCl caused the denatured proteins and the detergent to precipitate leaving the protein-free DNA (and RNA) in solution. The DNA that was lost from the soluble phase due to irradiation could be quantitatively accounted for in the precipitate containing the denatured proteins. Treatment of this material with trypsin, however, yielded free DNA. These data suggested that the DNA was cross-linked to protein. Further proof came from experiments showing that DNA and protein could be cross-linked *in vitro* (Figure 4-9).

The chemical mechanism(s) by which DNA and protein are cross-linked is not yet known, however, the isolation of a mixed photoproduct of uracil and cysteine (5-S-cysteine, 6-hydrouracil) (Figure 4-10) from the *in vitro* UV irradiation of a solution of uracil and cysteine may serve as a possible model for the cross-linking phenomenon. Cysteine-[35]S adds photochemically to poly-rU, poly-rC, poly-dC, poly-dT and to RNA and DNA. A mixed photoproduct of thymine and cysteine has been isolated. It has been identified as 5-S-cysteine, 6-hydrothymine. The presence of cysteine markedly inhibits the *in vitro* photochemical cross-linking of E. coli DNA and bovine serum albumin presumably by competing with the amino acid residues on the protein for attachment to the cytosine and thymine residues of the DNA. It has also been demonstrated that tyrosine and serine will add photochemically to DNA.

In addition to cysteine the following amino acids are also known to add photochemically to uracil: serine, cystine, methionine, lysine,

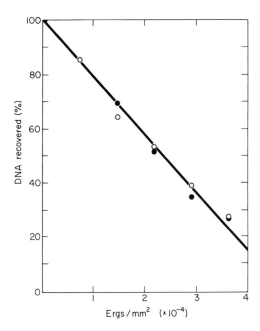

FIGURE 4-9. The photochemical cross-linking of DNA and protein *in vitro*. Solutions containing 10 mg bovine serum albumin and 0.03 mg DNA-thymine-2-$^{14}$C ($8.85 \times 10^4$ cpm/$A_{260}$ unit) in 4.2 ml $H_2O$ were irradiated (2537 Å) for various times and aliquots were processed for the recovery of DNA. There was no loss of DNA when it was irradiated in the absence of protein. [Modified from K. C. Smith, in "Radiation Research," p. 756 (G. Silini, ed.) Wiley, New York (1967).]

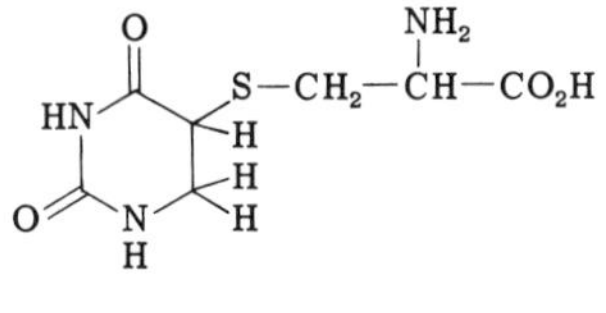

FIGURE 4-10. 5-*S*-cysteine, 6-hydrouracil. [From K. C. Smith and R. T. Aplin, *Biochemistry* 5, 2125 (1966).]

arginine, histidine, tryptophan, phenylalanine and tyrosine. The other common amino acids were unreactive under the conditions tested.

The biological importance of this DNA-protein cross-linking phenomenon has been indicated by studies in which the intrinsic sensitivity of bacterial cells to killing by UV has been changed by growth in nutritionally deficient media, and these changes have been accompanied by similar changes in the intrinsic sensitivity of the DNA to become cross-linked to protein by a constant dose of UV. The near equivalence in the timing of these changes in the sensitivity to killing and in the cross-linking phenomenon suggest that the cross-linking of DNA and protein may play a significant role in the loss of viability of these irradiated cells.

The experiments which best demonstrate the biological importance of DNA-protein cross-linking are those using bacterial cells irradiated while frozen. *Escherichia coli* B/r, T⁻ showed differences in survival after ultraviolet irradiation as a function of the temperature at which the cells were irradiated (Figure 4-11). When the temperature was reduced from +21° to −79°C, an increase in sensitivity to UV radiation was shown both by a change in extrapolation number (from 4 to 1) and a change in slope in the survival curves [$D_{37}$ (dose for 37% survival), 198 ergs/mm² at +21°C and 129 ergs/mm² at −79°C]. At −196°C the cells were not as sensitive as at −79°C ($D_{37}$, 198 ergs/mm²), but were more sensitive than at +21°C due to the absence of a shoulder.

A larger percentage of DNA was cross-linked to protein by a given dose of ultraviolet radiation when the cells were irradiated at −79°C or at −196°C as compared to +21°C (Figure 4-12). There is clearly a correlation in rank between the several cross-linking curves in Figure 4-12 and the survival curves in Figure 4-11.

In contrast, the rate of formation of cyclobutane-type thymine dimers decreased when the temperature of the cells during irradiation was varied from +21°C to −79°C and to −196°C (Figure 4-13). These curves show no correlation in rank with the survival curves in Figure 4-11.

Concomitant with this decrease in yield of thymine dimers in irradiated frozen cells, a decrease in the production of photoreactivable damage also occurred. This was seen when the cells at +21° and −79° were either exposed to the same dose of ultraviolet radiation or were killed to approximately the same survival value (Figure 4-14). Since the photoreactivating enzyme appears to be specific for the repair of cyclobutane-type pyrimidine dimers, the reduced amount of photoreactivation is consistent with the decrease in the production of thymine dimers under these conditions.

There is no correlation between the production of thymine dimers

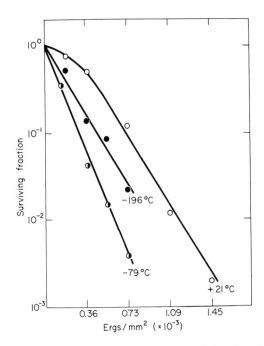

FIGURE 4-11.    Survival of *E. coli* B/r,T⁻ as a function of ultraviolet dose (2537 Å) at different temperatures. [From K. C. Smith and M. E. O'Leary, *Science* **155**, 1024 (1967).]

and the increased killing of *E. coli* by irradiation at −79° and −196°C. This suggests that cyclobutane-type thymine dimers do not play as significant a role in the events leading to the death of irradiated frozen cells as they appear to play at room temperature. These results provide further evidence that the relative biological importance of a given photoproduct can change markedly, depending upon growth or irradiation conditions.

The photochemical event that does correlate with viability when cells are irradiated while frozen is the cross-linking of DNA with protein. Freezing produces both a change in the rate of formation and in the yield of DNA cross-linked to protein. Freezing may alter the configuration or the proximity of the protein and the DNA within the cells so that the probability of forming DNA-protein cross-links by irradiation is greatly enhanced, thus leading to the greater lethality observed under these conditions.

The action spectrum for the killing (and for the inhibition of DNA

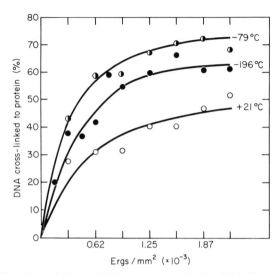

FIGURE 4-12.   Cross-linking of DNA and protein in *E. coli* B/r,T⁻ as a function of
ultraviolet dose (2537 Å) at different temperatures. The values plotted are the average values for five experiments at +21°C and two each at −79°C and −196°C. [From
K. C. Smith and M. E. O'Leary, *Science* **155**, 1024 (1967).]

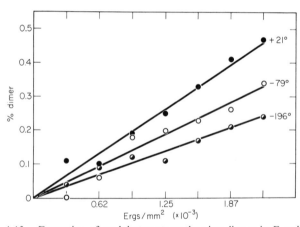

FIGURE 4-13.   Formation of cyclobutane-type thymine dimers in *E. coli* B/r,T⁻ as
a function of ultraviolet dose (2537 Å) at different temperatures. Cells labeled with
thymine-2-¹⁴C were irradiated, hydrolyzed in trifluoroacetic acid, and chromato-
graphed. The cells used were from certain of the experiments described in Figure
4-12. These results are the average of two experiments at +21°C and one each at the
other temperatures. [From K. C. Smith and M. E. O'Leary, *Science* **155**, 1024
(1967).]

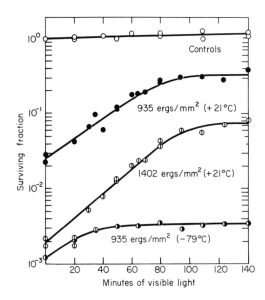

FIGURE 4-14.  Photoreactivability of *E. coli* B/r,T⁻ after ultraviolet irradiation (2537 Å) at +21°C and −79°C. Stationary cultures were suspended (5 × 10⁸ cell/ml) in 0.1$M$ phosphate buffer at pH 6.8. The suspension was frozen at −79°C for 30 minutes and irradiated, or thawed and irradiated at +21°C. Samples on a shaker table were then exposed to visible light at +21°C for the times indicated, and viability was determined on plates of nutrient agar. The visible light source was two 15-watt Westinghouse Daylight fluorescent bulbs 9 cm above the bottom of the Petri dish. The top half of the glass Petri dish was inverted, filled with 15 ml of water, and placed over the bottom half of the Petri dish to protect the cells therein from UV radiation and heat. [From K. C. Smith and M. E. O'Leary, *Science* **155**, 1024 (1967).]

synthesis) of *Micrococcus radiodurans*, one of the most radiation resistant organisms known, differs markedly from that for the more sensitive organism *E. coli* in that it shows a high component of sensitivity to irradiation at 2800 Å as well as at 2600 Å. Classically, a response at 2800 Å has indicated an involvement of protein (Section 1-4). It has been suggested that the resistance of this organism to UV is due to its extraordinary ability to repair thymine dimers, but that what ultimately kills the organism is damage that involves both DNA and protein. The cross-linking of DNA and protein may constitute one type of such damage.

It is reasonable to assume that a different type of photochemistry might arise when protein and DNA are irradiated together as compared to when they are irradiated separately. Since DNA and protein do not exist in cells as pure solutions of the separate molecules but are in intimate contact with each other, it might be expected that the photochemical interaction of DNA and protein would play a significant role in the inactivation of UV-irradiated cells under certain conditions.

4-10. The Effect of Base Composition on the Intrinsic Sensitivity of DNA to UV

Separate genetic markers of transforming DNA with different base compositions have been shown to differ in their sensitivity to UV inactivation. This result is in harmony with the observation that there is a correlation between the base composition of the DNA of a particular bacterial strain and its radiation sensitivity. Thus, as the adenine-thymine content increased (guanine-cytosine content decreased) the cells showed an increased sensitivity to killing by UV. This relationship would seem to be adequately explained by our present knowledge of the importance of thymine photoproducts in the UV inactivation of DNA.

4-11. The Effect of Substitution by Halogenated Pyrimidines on the Intrinsic Sensitivity of DNA to UV

The van der Waals radii of the chloro, iodo, and bromo groups are similar to that of the methyl group of thymine (Figure 4-15). Therefore, these 5-halogenated derivatives of uracil are mistaken for thymine by certain enzymes and are consequently incorporated into DNA in place of thymine. The van der Waals radius of the fluoro group is similar to that of hydrogen so that 5-fluorodeoxyuridine is an analog of deoxyuridine. It inhibits DNA synthesis by blocking the methylation of deoxyuridylic acid to form thymidylic acid. If fluorouracil is combined with ribose it will be incorporated into RNA in place of uracil.

RNA viruses (e.g., tobacco mosaic virus) are somewhat more sensitive to UV inactivation when part of their uracil is replaced by 5-fluorouracil. Bacteria or mammalian cells that have incorporated halogenated analogs of thymine into their DNA are much

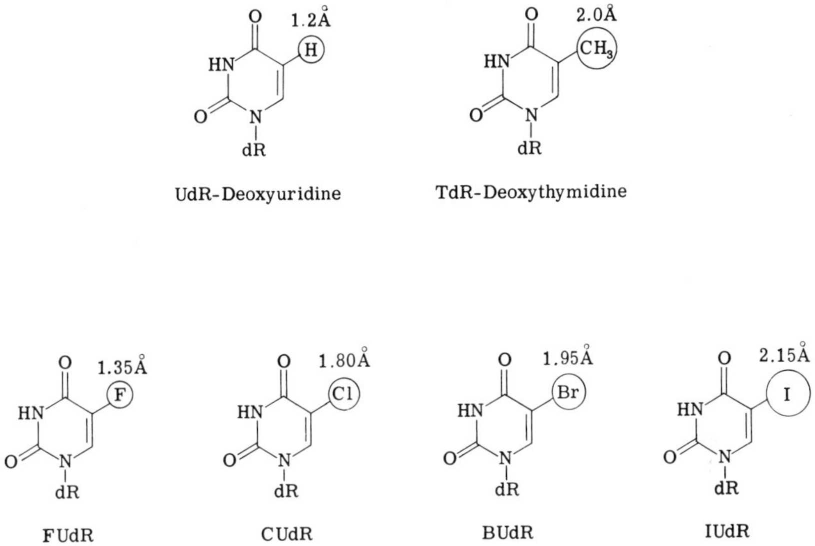

UdR-Deoxyuridine          TdR-Deoxythymidine

FUdR          CUdR          BUdR          IUdR

FIGURE 4-15. Chemical structure of natural deoxynucleosides, deoxyuridine (UdR) and thymidine (TdR), and of their halogenated analogs: 5-fluorodeoxyuridine (FUdR), 5-chlorodeoxyuridine (CUdR), 5-bromodeoxyuridine (BUdR), and 5-iododeoxyuridine (IUdR). The circles in position 5 represent the comparative sizes of the substituting group, the van der Waals radii of which are indicated above each circle. Because of the similarity of the van der Waals radii, FUdR behaves as an analog of deoxyuridine, whereas CUdR, BUdR and IUdR are analogs of thymidine. [From W. Szybalski, in "The Molecular Basis of Neoplasia," p. 147. Univ. of Texas Press, Austin, Texas, 1962.]

more sensitive to killing by UV (Figure 4-16). For a given dose of UV, the 5BU in the DNA of bacterial cells is about twice as sensitive to photochemical alteration as is thymine and a greater number of different photoproducts are formed relative to thymine. This alteration in the intrinsic photochemical sensitivity of the DNA must certainly contribute significantly to the altered radiation sensitivity of these cells.

Cells containing 5BU-substituted DNA show a fivefold greater sensitivity to UV-induced DNA-protein cross-linking than normal cells. Irradiated DNA containing 5BU exhibits a greater sensitivity to intramolecular-interstrand cross-linking, and is more susceptible to a UV-induced decrease in molecular weight.

The increase in photochemical sensitivity of DNA by analog substitution, however, is not the only manifestation of the altered radiation sensitivity of analog substituted cells. Irradiated bacteriophage that are substituted with 5BU cannot be photoreactivated nor can they be dark reactivated. The photoproducts of 5BU therefore do not seem to be amenable to repair by currently known mechanisms.

The increased sensitivity of halogenated-pyrimidine-substituted-DNA *in vivo* and *in vitro* can be explained both on the basis that these analogs show an increased sensitivity to photochemical alteration and by the fact that the photoproducts produced seem to be refractory to repair.

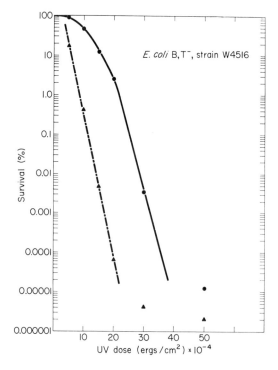

FIGURE 4-16.   The effect of ultraviolet irradiation on the survival curves of *E. coli* B.T⁻, grown on thymine (●) or 5-bromouracil (▲). [From H. S. Kaplan, K. C. Smith, and P. A. Tomlin, *Radiation Res.* 16, 98 (1962).]

4-12. THE INFLUENCE OF THE ENVIRONMENT DURING IRRADIA-
TION ON THE INTRINSIC SENSITIVITY OF DNA

The photochemical reactivity of thymine is markedly different if it is irradiated in solution, in frozen solution, or in dry films (Section 4-5). 5-Bromouracil is inert to UV if irradiated in frozen solution unless additional compounds are added.

This extreme importance of the environment on the photochemistry of the bases is also carried over to the photochemistry of DNA and to the sensitivity of cells to irradiation. The formation of the cyclobutane-type thymine dimer is greatly depressed if the DNA is irradiated dry. If bacteriophage T1 are UV-irradiated while dry they cannot be photoreactivated. Little or no cyclobutane-type thymine dimers are formed in irradiated spores, yet they are the major photoproduct produced in irradiated vegetative cells. The DNA within spores is thought to be dry and spores are more resistant to killing by UV than are vegetative cells. Cells that have been irradiated while frozen are much more sensitive to killing by UV. Under these conditions the yield of thymine dimers is greatly depressed but there is an increase in the amount of DNA cross-linked to protein.

Coupled with the effects of the environment on the photochemistry of DNA are the effects of the physical state of the DNA itself upon its susceptibility to UV alteration. The rate of thymine dimer formation in denatured DNA, for example, is about twice that for native DNA, although the yield is the same. The formation of hydrates is also greatly favored in single-stranded DNA.

4-13. PHOTOCHEMISTRY OF RNA

There is not as much information available on the photochemistry of RNA as there is for DNA; however, there have been several recent publications by H. E. Johns and co-workers on the factors affecting the rate of hydrate and dimer formation in various single and double-stranded polymers containing uridylic acid and/or cytidylic acid. The information cited previously (Section 4-4) on the effects of UV on the coding properties of poly-rC in an RNA polymerase system is also pertinent.

Uracil dimers have been isolated from irradiated RNA. The

photoreactivability of tobacco mosaic virus (TMV) RNA in the tobacco leaf is suggestive but certainly does not prove that pyrimidine dimers are inactivating lesions. In fact, current evidence suggests that hydrates rather than dimers may be the lesions responsible for inactivation. The extent of photoreactivation has been found to be independent of the wavelength of inactivation in the range of 2300 Å to 2800 Å. TMV-RNA irradiated in the region of 3000 Å, however, exhibited a greater degree of photoreactivation than that irradiated at 2537 Å. The explanation of these results awaits the identification of the photoproduct whose production is constant between 2300–2800 Å but is markedly enhanced at 3000 Å. There is some preliminary evidence to suggest that mechanisms are present in plants for the dark repair of UV-irradiated RNA, but further work on this subject is needed.

Whole irradiated TMV are not photoreactivable, suggesting that the lesions leading to the inactivation of the virus may be different from those leading to the inactivation of irradiated extracted viral RNA. In this regard, it is interesting that the cross-linking of RNA and protein has been observed in irradiated TMV. The inactivation of a virus particle occurs concomitantly with the binding of approximately one protein subunit to the RNA of the virus. The mechanism of this response may be similar to that involved in the cross-linking of DNA and protein in irradiated bacteria (Section 4-9).

The quantum yield (2537 Å) for the inactivation of transfer RNA from mouse liver is $6.7 \times 10^{-4}$. Various transfer RNA's from $E.$ $coli,$ yeast, and mouse liver show wide differences in sensitivity to UV as measured by the rate of inactivation of their amino acid acceptor capacities. Reirradiation at short wavelengths and heating are not effective in restoring acceptor activity. The cause of inactivation would seem to be a change in the secondary structure of the transfer RNA brought about by the production of photoproducts other than uracil dimers or hydrates. The $in\ vivo$ inactivation by UV of the messenger RNA for an induced enzyme in $E.\ coli$ has also been reported.

4-14. SUMMARY

Some of the biological effects of UV irradiation can now be explained in terms of specific chemical and physical changes produced in DNA.

Photochemical events in the carbohydrate and purine residues of DNA would appear to occur too infrequently to be of major biological importance, but this point has not been adequately investigated.

Pyrimidine hydrates do not seem to be formed efficiently in double-stranded DNA but are formed in single-stranded DNA. The possible importance of hydrates in causing mutations has been demonstrated.

Cyclobutane-type dimers are formed by the pyrimidines, separately and as mixed dimers. The biological importance of this type of dimer has been demonstrated in certain situations. This type of photoproduct is not formed in DNA under all conditions so it cannot occupy a position of supreme biological importance in all situations. Other types of photoproducts therefore must also be of significance.

The biological importance of DNA-protein cross-linking at low doses of UV has been demonstrated under certain conditions. One chemical mechanism for this cross-linking may involve the attachment of amino acid residues through their SH (or OH) groups to the 5 (or 6) carbon of cytosine and thymine.

UV irradiation causes chain breakage and the formation of DNA-DNA cross-links but these usually occur only at high doses so that their importance at low doses seems questionable.

The intrinsic sensitivity of DNA to alteration by UV can be effected by a change in base composition, substitution by analogs, and by altering the environment or physical state of the DNA during irradiation.

Hydrates and dimers have been observed in irradiated RNA, and RNA-protein cross-linking has been observed in irradiated tobacco mosaic viruses (TMV). TMV cannot be photoreactivated on the plant leaf but irradiated TMV–RNA can. In the latter case it appears that hydrates rather than dimers are the cause of inactivation of the RNA. RNA protein cross-links appear to be involved in the inactivation of the whole virus.

Although a given photochemical lesion has been shown to be of biological importance under certain conditions it is not expected that it should enjoy similar importance under all conditions. The biological importance of any photoproduct depends upon (1) whether or not it is formed under a particular set of experimental conditions and (2) if formed, whether or not the particular system under study is capable of repairing the lesion.

GENERAL REFERENCES

D. Shugar, Photochemistry of nucleic acids and their constituents. *In* "The Nucleic Acids" (E. Chargaff and J. N. Davidson, eds.), Vol. 3, pp. 39–104. Academic Press, New York, 1960.

A. Wacker, Molecular mechanisms of radiation effects. *Prog. Nucleic Acid. Res.* 1, 369–399 (1963).

K. C. Smith, Photochemistry of nucleic acids. *In* "Photophysiology" (A. C. Giese, ed.), Vol. 2, pp. 329–388. Academic Press, New York, 1964.

R. B. Setlow, Molecular changes responsible for ultraviolet inactivation of the biological activity of DNA. *In* "Mammalian Cytogenetics and Related Problems in Radiobiology" (C. Pavan *et al.,* eds.), pp. 291–307. Pergamon Press, Oxford, 1964.

A. D. McLaren and D. Shugar, "Photochemistry of Proteins and Nucleic Acids." Pergamon Press, Oxford, 1964.

K. C. Smith, Physical and chemical changes induced in nucleic acids by ultraviolet light. *Radiation Res.,* Suppl. 6, 54–79 (1966).

J. K. Setlow, The molecular basis of biological effects of ultraviolet radiation and photoreactivation. *Current Topics Radiation Res.,* 2, 195–248 (1966).

J. G. Burr, *Advan. Photochem.* 6, 193–299 (1968).

# 5

# Photochemistry of
# Amino Acids and Proteins

## 5-1.  INTRODUCTION

During the past 10–15 years, general interest in photobiology has strongly emphasized the nucleic acids, although currently there is renewed interest in photon effects on proteins. Most of the data on the photochemical reactions of the amino acids, however, were gathered some years ago without the benefit of monochromatic light sources, the availability of labeled amino acids, or of modern methods of chromatography and electrophoresis for the separation of the photoproducts. The reported quantum yields for proteins are sometimes in error due to inadequate knowledge (at that time) of the molecular weight and amino acid composition of the particular proteins under investigation. The ready availability of synthetic polyamino acids now provides an excellent opportunity for further exciting work on the photochemistry of the amino acids especially as regards possible energy migration and neighbor interactions. Our understanding of the effects of UV on cells requires an understanding of the chemical modifications produced in proteins by UV. The finding that proteins and DNA interact photochemically within the cell (Section 4-9) further necessitates a renaissance in studies on the photochemistry of amino acids and proteins.

85

5-2. RELATIVE PHOTOCHEMICAL SENSITIVITY OF THE AMINO
ACIDS

Since light must be absorbed before photochemistry will occur, it
is reasonable to assume that those amino acids that are transparent
to the wavelengths normally used in photobiology (wavelengths
greater than 2400 Å) will undergo little or no photochemistry at
these wavelengths. This then largely eliminates the aliphatic amino
acids from our consideration. The strong absorption of wavelengths
above 2400 Å by the aromatic amino acids would certainly suggest
their photochemical importance, although it should be recalled that
absorbed light need not necessarily result in photochemistry (Section 1-2).

Before continuing with the photochemistry of the aromatic amino
acids, let us consider two chromophores that are not present in
monomeric amino acids but are present in proteins: peptide linkages
and disulfide bridges. The absorption spectrum for peptide bonds
shows a peak in the region of 1800–1900 Å which decreases essentially to zero by about 2400 Å. Peptide bonds are broken by absorption at these wavelengths. Therefore, if peptide bond breakage
is to be minimized, wavelengths below 2400 Å must be filtered out
of the light source. Even at 2537 Å, however, the quantum yield
for peptide bond breakage is a significant value (Table 5-1).

The amino acids, cysteine and cystine, were considered at one
time to be unimportant photochemically since according to their
absorption spectra in acid solution they were essentially optically
transparent above 2300 Å. At neutral and alkaline pH, however,
the absorption by these compounds becomes appreciable even at
wavelengths above 2500 Å. This fact coupled with the high quantum
efficiency for their photochemical alteration makes cystine the most
sensitive target affecting protein function.

If the molar absorptivity ($\epsilon$) of a compound is a measure of the
probability that light of a particular wavelength will be absorbed by
that compound, and the quantum yield ($\Phi$) is the probability that the
absorbed light will cause a chemical change, then the product of
these two values ($\epsilon \times \Phi$) is a measure of the photochemical sensitivity of that compound. Values for these parameters at 2537 Å for
the most important chromophores in proteins are given in Table 5-1.
Thus, when cystine is present in a protein it is the most labile target.
Following in decreasing lability are tryptophan, phenylalanine,

TABLE 5-1. The Relative Photochemical Lability of Amino Acids at 2537 Å[a]

| Compound | $\epsilon^b$ | $\Phi^c$ | $(\epsilon \times \Phi)^d$ |
|---|---|---|---|
| Cystine (—S—S— bond) | 270 | 0.13 | 35.1 |
| Tryptophan | 2870 | 0.004 | 11.5 |
| Phenylalanine | 140 | 0.013 | 1.8 |
| Tyrosine | 320 | 0.002 | 0.6 |
| Peptide bonds (acetylalanine) | 0.2 | 0.05 | 0.01 |
| Histidine | 0.24 | <0.03 | <0.0072 |

[a]Data from A. D. McLaren and D. Shugar, "Photochemistry of Proteins and Nucleic Acids," p. 97. Pergamon Press, Oxford, 1964.

[b]$\epsilon$ indicates molar absorptivity at 2537 Å, or the probability that light at 2537 Å will be absorbed by the given compound.

[c]$\Phi$ indicates quantum yield (the probability that the absorbed light (2537 Å) will cause a chemical change).

[d]$(\epsilon \times \Phi)$ indicates a measure of the photochemical sensitivity of a particular amino acid (it is related to the inactivation cross-section, $\sigma$, by a constant; Section 6-2).

tyrosine, peptide bonds, and histidine. It should be stressed, however, that at other wavelengths the UV sensitivity and therefore the relative importance of these targets in the photochemical inactivation of a particular protein will change.

Most of the early photochemical studies on the amino acids did not use monochromatic light and the analysis for photochemical alteration relied upon easily measurable products such as ammonia, carbon dioxide, acids, bases, sulfur and loss of spectral properties. The importance of using monochromatic light for photochemical studies has already been stressed (e.g., Section 4-5). Relying only upon end products of destruction such as ammonia, carbon dioxide, etc., very little information can be gained about intermediate reactions that occur at biologically significant doses of UV.

A recent study on the photochemistry of phenylalanine (selectively labeled at different positions with [14]C) has shown that when this amino acid is irradiated at the wavelength of maximum absorption of the aromatic residue (2575 Å) that tyrosine, dihydroxyphenylalanine (DOPA), aspartic acid, benzoic acid, phenyllactic acid, and a high molecular weight melanic polymer are formed. The reaction that occurred with the greatest quantum efficiency was the liberation of carbon dioxide by decarboxylation. These results indicate that there was cleavage of side chain carbon bonds, and since all of the energy was initially absorbed by the benzene ring, it is therefore reasonable to conclude that energy migration by some mechanism (as yet not understood) must have occurred. The direct transmission of energy along —$(CH_2)_n$— would appear to be inconsistent with the observation (Section 1-3) that methylene groups affect the absorption spectra of compounds by acting as insulators between aromatic groups.

Several other examples of energy migration in peptides and proteins are known. The photochemical decomposition of carbon monoxide–myoglobin occurs with the same quantum yield regardless of whether the light energy is absorbed by the aromatic amino acids of the protein or the porphyrin ring. When the dye 1-dimethylaminonaphthalene-5-sulfonyl chloride was bound to proteins and the complex irradiated with light that was specifically absorbed by the aromatic residues of the protein it was found that the dye fluoresced. The fluorescence of the dye could only be explained on the basis of energy transfer from the protein. This type of response has also been shown to result in a decrease in the inactivation rate of irradiated proteins. Energy migration is thought to occur in part through a *resonance-transfer mechanism*. A fluorescent donor, with a fluorescence spectrum overlapping the absorption spectrum of the energy acceptor, can pass energy to the acceptor if the distance between the two is small (less than 100 Å) and if the relative orientation between the two is suitable (Section 3-7). Since it is possible to distinguish between the spectra of phenylalanine, tyrosine and tryptophan the fluorescence and phosphorescence spectra of proteins can yield information on the direction of transfer of energy between these amino acids. Phenylalanine fluorescence is not observed from most proteins although a tyrosine component is always observed. Tyrosine fluorescence makes only a small contribution to the total fluorescence from those proteins containing tryptophan. The yields of cystine destruction are much higher in those proteins that contain tryptophan, again emphasizing the importance of energy transfer in an irradiated protein (Section 5-4).

5-3. GENERAL COMMENTS ON THE PHOTOCHEMISTRY OF PROTEINS

(a) When proteins are irradiated with UV, both lower and higher molecular weight products are formed. The higher molecular weight aggregates are much easier to detect, however, and appear to be produced with the greatest efficiency. While aggregation proceeds at the same rate under oxygen or nitrogen at lower doses of UV, at higher doses in the presence of oxygen, the aggregates begin to decompose and the solubility of the irradiated protein therefore begins to increase.

(b) Although certain aspects of the photochemistry of proteins are dependent upon whether they are irradiated in the presence of oxygen (see above), the quantum yield for inactivation of an enzyme does *not* depend upon the atmosphere during irradiation.

(c) Irradiated proteins are much more susceptible to enzymatic digestion. This, coupled with the fact that heat denatured proteins are in general more susceptible to enzymatic digestion, indicates that UV causes the denaturation of proteins. It has also been observed that following doses of UV (which cause no serious loss in enzymatic activity) the enzymes are much more sensitive to subsequent heat inactivation. At least two mechanisms could explain this observation. If a peptide bond were broken but the broken ends were held in place by the hydrogen bonds involved in the secondary and tertiary structures of the protein, then the molecule would be enzymatically active until gently heat treated. If a disulfide bridge were broken by UV, then it might require less heat to disorganize the hydrogen bonds of the molecule and in the absence of the directional influence of the disulfide bridge, the original hydrogen bonds would be less likely to reform on cooling than they would have prior to irradiation.

(d) Near room temperature, the quantum yield for the inactivation of an enzyme does not show any marked dependency upon the temperature during irradiation. However, at temperatures below room temperature there is a continuous decrease (as the temperature decreases) in the quantum yield; while as the temperature increases above room temperature, there is a continuous increase in quantum yield. We have commented previously on the effect of temperature and environment upon the absorbance of various compounds (Chapter 1) and on the photochemical reactivity of DNA (Chapter 4). Such considerations may possibly be invoked to explain the effect of temperature on the photochemical reactivity of proteins.

(e) The quantum yields for the UV inactivation of enzymes irradiated in the dry state and in aqueous medium do not differ significantly. This suggests that the mechanisms involved in the two cases are not radically different, but this may well be an over-simplification.

(f) The quantum yield for proteins varies with the wavelength of irradiation. The explanation for this is that the action spectrum for the inactivation of an enzyme can be resolved into the contributions by the various chromophores. At 2537 Å, the quantum yield for the inactivation of trypsin can be adequately accounted for by assigning the major role to cystine, but at wavelengths above this value the aromatic amino acids play a more important role and at wavelengths below 2537 Å peptide bonds play an important role.

(g) The quantum yield (at 2537 Å) for the inactivation of an enzyme is roughly proportional to its cystine content. This is consistent with the data cited in Table 5-1 and will be more fully discussed in Section 5-4.

(h) The quantum yield for the inactivation of an enzyme is roughly

TABLE 5-2. SOME CHANGES PRODUCED IN PROTEINS FOLLOWING EXPOSURE TO UV[a]

| Property | Change | Possible meaning |
|---|---|---|
| Color | Darkens | Photooxidation |
| Solubility in water | Decreases; then increases | Aggregation, then degradation |
| Sensitivity to heat | Usually increases | Hydrophobic groups exposed; chain breaks |
| Optical rotation | Increase in levorotation | Structural changes |
| Change in absorption | Increases | Depolymerization and photooxidation; formation of new chromophores |
| Digestibility by trypsin | Increases | Opening up of molecule |
| Exposure of SH groups | Increases | Opening up of molecule |
| Exposure of acid groups | Increase; pH change | Photooxidation to form more phenols; breakage of peptide bonds |
| Electrophoretic pattern | Heterogenization | Degradation and reaggregation |
| Ultracentrifugation | Increased broadening of sedimentation bands | Molecular weight heterogeneity |
| Immunological effects | Destruction of antigenic properties; formation of new ones | Altered amino acid residues |

[a]Modified from A. C. Giese, *Physiol. Rev.* **30**, 431 (1950).

proportional to the reciprocal of its molecular weight. A precise explanation for this observation is not available, however, it has been pointed out that the cystine content of proteins is roughly proportional to the reciprocal of their molecular weights.

(i) In most cases the quantum yields for the inactivation of the nucleic acids are less than those for proteins, yet the molar absorbtivity for the proteins are much lower than for the nucleic acids. McLaren has observed that the product of the molar absorptivity ($\epsilon$) and quantum yield ($\Phi$) for proteins at 2537 Å is approximately equal to that for the nucleic acids. The significance of this observation remains to be evaluated but at face value it would imply that proteins may play a larger role in biological inactivation than most of the nucleic acid chemists would care to admit.

(j) The quantum yields for the inactivation of enzymes usually vary with pH but there appears to be no correlation with the isoelectric point of the enzyme. The overall charge on the molecule would therefore appear to be of less importance than the charge on some specific amino acid residue(s).

(k) Other changes noted in proteins following exposure to UV are listed in Table 5-2.

## 5-4. PHOTOCHEMICAL INACTIVATION OF ENZYMES

Two different theories of the mechanisms involved in the UV inactivation of enzymes have developed over a period of years. McLaren and co-workers have assumed, as a first approximation, that the alteration of any amino acid residue causes inactivation of an enzyme. If the quantity ($\epsilon \times \Phi$) for each amino acid (which is a measure of the photochemical sensitivity of each amino acid; see Table 5-1) is multiplied by the number of such amino acids (n) in the protein and the sum of these is divided by the molar absorptivity of the enzyme, one arrives at a calculated quantum yield for the inactivation of the enzyme which in some cases is very close to the observed quantum yield. Of the ten proteins listed in Table 5-3, the calculated quantum yields for seven of these are within a factor of 2 of the experimentally determined values. The values for carboxypeptidase, pepsin, and insulin do not adequately fit the hypothesis.

The authors themselves have pointed up certain criticisms of these calculations. There is no assurance that either the molar absorptivity or the quantum yield for a particular amino acid (at a given pH and wavelength) is the same in a protein as it is in free solution; in fact, there is some evidence that these values change when the amino acid is incorporated into a protein. Also it is known that all residues of a particular amino acid in a given protein are *not* of equal importance to enzymic function.

For those few proteins studied, their absorbance at 2537 Å does not differ markedly from that for an equivalent mixture of the free amino acids although greater differences exist at other wavelengths. This fact, plus the rather close agreement between the calculated and determined quantum yields cited in Table 5-3 would suggest that under certain conditions the amino acid residues are, in fact, independent absorbers unaffected by their neighbors. McLaren and co-workers therefore state that energy transfer among chromophores within molecules of enzymes need not be invoked in order to account for photochemical inactivation.

Augenstein and associates, however, take particular exception to this conclusion. They have developed a theory that enzymes are inactivated by UV as a consequence of the disruption of specific cystine residues and of hydrogen bonds responsible for the spatial integrity of the active center of the enzyme. Their theory emphasizes the importance of energy migration in this process. Consistent with these postulates are the observations that the photochemical damage in the enzyme ribonuclease is nonrandom; at least two and perhaps three of the four constituent cystines must be disrupted before activity is lost, i.e., the most photosensitive cystines are not critical for enzymic activity. Similarly, in both trypsin and lysozyme the integrity of the most photosensitive cystines does not appear to be critical for the retention of enzymic potential. In insulin, however, all three cystines appear to be crucial for activity and have approximately equal photosensitivities. These differences in sensitivity of cystines in different proteins must depend specifically upon energy transfer and/or chemical interactions between the chromophoric groups. If quantum yields are calculated on the basis of those quanta absorbed only in the cystines, values for cystine destruction in proteins are obtained which are about 5 to 8 times greater than those observed in the model compounds, cystine and oxidized glutathione. Cystine destruction is much greater in those proteins which contain tryptophan. There is some evidence that the most photosensitive cystines lie close to tryptophan residues as a consequence of the tertiary structure of the enzyme rather than of its primary structure. We seem to be faced with two opposing theories concerning the involvement of energy transfer in the UV inactivation of enzymes.

TABLE 5-3. CALCULATED AND KNOWN QUANTUM YIELDS FOR ENZYME INACTIVATION BY UV (2537 Å)[a]

| | | Quantum yield | |
|---|---|---|---|
| Protein | Chromophores | Found | Calculated[b] |
| 1. Carboxypeptidase | $Cys_2 \cdot His_8 \cdot Phe_{15} \cdot Try_6 \cdot Tyr_{20}$ | 0.001[c] | 0.01 |
| 2. Chymotrypsin | $Cys_5 \cdot His_2 \cdot Phe_6 \cdot Try_7 \cdot Tyr_4$ | 0.005 | 0.01 |
| | | | $(0.007)^d$ |
| 3. Lysozyme | $Cys_5 \cdot His_1 \cdot Phe_3 \cdot Try_8 \cdot Tyr_2$ | 0.024 | 0.014 |
| | | | $(0.0097)^d$ |
| 4. Pepsin | $Cys_3 \cdot His_2 \cdot Phe_9 \cdot Try_4 \cdot Tyr_{16}$ | 0.002 | 0.01 |
| | | | $(0.006)^d$ |
| 5. Ribonuclease | $Cys_4 \cdot His_4 \cdot Phe_3 \cdot Try_0 \cdot Tyr_{16}$ | 0.027 | 0.03 |
| | | | $(0.026)^d$ |
| 6. Subtilisin A | $Cys_0 \cdot His_5 \cdot Phe_4 \cdot Try_1 \cdot Tyr_{13}$ | 0.007 | 0.0051 |
| 7. Subtilisin B | $Cys_0 \cdot His_6 \cdot Phe_3 \cdot Try_4 \cdot Tyr_{10}$ | 0.006 | 0.011 |
| 8. Japanese Nagarse | $Cys_0 \cdot His_6 \cdot Phe_3 \cdot Try_4 \cdot Tyr_{10}$ | 0.007 | 0.01 |
| 9. Trypsin | $Cys_6 \cdot His_1 \cdot Phe_3 \cdot Try_4 \cdot Tyr_4$ | 0.015 | 0.02 |
| | | | $(0.014)^d$ |
| 10. Insulin | $Cys_{18} \cdot His_{12} \cdot Phe_{18} \cdot Try_0 \cdot Tyr_{24}$ | 0.015 | 0.06 |

[a]From A. D. McLaren and O. Hidalgo-Salvatierra, *Photochem. Photobiol.* 3, 349 (1964).

[b]$\Phi_{ENZ} = \overset{i}{\Sigma} \eta_i \, \epsilon_i \, \Phi_i / \epsilon_{ENZ}$     (see Table 5-1 and Section 5-4).

[c]The quantum yield for carboxypeptidase has been redetermined by R. Piras and B. L. Vallee [*Biochemistry* 6, 2269 (1967)] to be 0.0049 which now brings the calculated value (0.01) within about a factor of two of the observed value. Newer data (cited by Piras and Vallee) indicate that carboxypeptidase contains no S—S bridges.

[d]Often a much closer agreement is obtained between the observed and calculated quantum yields if only the cystine residues are used in the calculations instead of using all of the amino acids that absorb UV light. This is permissible as a special condition of the formulation but suggests that the statement by McLaren and Hidalgo-Salvatierra (citation in footnote a) that "the site of quantum absorption and the site of photochemical reaction in proteins are one and the same and that it is not necessary to invoke a mechanism of migration of absorbed energy from one kind of chromophore to another within an absorbing molecule" is not valid for all proteins. for all proteins.

One may take exception with certain of the assumptions used by McLaren and co-workers in developing their theory but these probably would not cause order of magnitude (ten fold) errors. The calculated quantum yields for seven enzymes agree with the observed quantum yields within a factor of two and this must be considered to support their theory. However, the danger of theories is that they may be accepted as universal and therefore stifle future research. Perhaps an equally significant contribution that this theory can make to the photochemistry of proteins is the fact that it does have several marked exceptions. Augenstein's theory deals principally with those proteins which do not closely fit the theory of McLaren, and the work of Augenstein and co-workers offers convincing data in support of the importance of energy migration in the photochemical inactivation of certain proteins.

We may conclude that both theories are correct but for different proteins. There must be unique structural features or unique amino acid sequences in some proteins that allow the amino acids to be independent absorbers and for photochemical inactivation events to occur apparently at random, while different structural features in other proteins give rise to the opposite results (i.e., nonrandom photochemical events and the involvement of energy migration in the inactivation of enzymes).

Although it has long been known that UV has a deleterious effect upon proteins, we still have very little understanding of the fundamental chemical mechanism(s) involved. In fact, our knowledge concerning the photochemistry of the individual amino acids is grossly inadequate. Also, there is evidence that not all of the molecules of a particular amino acid within a protein have the same photochemical sensitivity. Until we have a better understanding of the primary photochemistry of the individual amino acids (singly and in peptides), we cannot hope to understand properly the photochemical mechanisms involved in the inactivation of proteins.

Interest in the photochemistry of the amino acids and proteins has been overshadowed in recent years by the phenomenal surge in research activity on the photochemistry of the nucleic acids. Our knowledge of the photochemistry of the nucleic acids now far exceeds our knowledge of the photochemistry of the proteins. An adequate knowledge of both is necessary, however, before we can properly understand the mechanism of action of UV on cells. The paucity of knowledge of the molecular mechanisms involved in the photochemistry of the proteins (and amino acids) makes it a very attractive field for future research.

## GENERAL REFERENCES

R. B. Setlow, A relation between cystine content and ultraviolet sensitivity of proteins. *Biochim. Biophys. Acta* 16, 444 (1955).

R. B. Setlow, Radiation studies on proteins and enzymes. *Ann. N.Y. Acad. Sci.* 59, 469 (1955).

A. D. McLaren and D. Shugar, "Photochemistry of Proteins and Nucleic Acids." Pergamon Press, Oxford, 1964.

A. D. McLaren and O. Hidalgo-Salvatierra, Quantum yields for enzyme inactivation and the amino acid composition of proteins. *Photochem. Photobiol.* 3, 349 (1964). Theory: Random destruction of amino acids causes inactivation. Energy transfer considerations are not necessary to explain inactivation.

L. Augenstein and P. Riley, The inactivation of enzymes by ultraviolet light. IV. The nature and involvement of cystine disruption. *Photochem. Photobiol.* 3, 353 (1964); see also, *ibid.* 6, 423 (1967).
Theory: Inactivation is caused by the disruption of specific cystine residues and of hydrogen bonds responsible for the spacial integrity of the active center of the enzyme. Energy transfer considerations are important.

# 6

# Photoinactivation of Biological Systems

## 6-1. INTRODUCTION

In previous chapters we have considered the known photochemical events that may contribute to the deleterious biological effects of UV photons. It is our purpose now to examine some of the classes of biological inactivation effects and their interpretations. As might be predicted, such interpretations become much more difficult when the sample is not merely a solution of DNA but is rather a metabolizing, living cell containing a variety of mechanisms specifically designed for dealing with possible inactivating events. We will not be able to assume that the expressed biological effect is the result of all of the initial photochemistry since many of the photoproducts may be rendered harmless by repair mechanisms. Also, it is important to realize that the metabolizing cell may exhibit different sensitivities to inactivating events at different stages in the growth cycle as the numbers and physical configurations of the pertinent macromolecules change.

It is not too surprising to find that DNA is the principal target for most of the deleterious effects of photons on growing cells. This is due to the ultimate importance of DNA to the duplication of the cell and to its sensitivity to alteration in terms of precision of function. Damage to a single nucleotide in DNA may be sufficient to kill a cell. On the other hand such damage to a nucleotide may result in a nonlethal mutation, or it may not be detectable by any biological assay. Under conditions of a multiplicity of genetic information (e.g., two nuclei in the same cell) damage in one genome may go undetected if the other nucleus is intact. A basic approach in photobiological studies with such complicated systems involves the determination of a radiation dose-effect curve for the effect of interest. The dependence of the measured effect (e.g., viability) upon increasing dose of radiation not only provides one with an indication of the sensitivity of the system but it also is a first step in determining which parameters are involved. A large part of UV photobiology is thus concerned with the interpretation of survival curves.

6-2.   SIMPLE SURVIVAL CURVES

By analogy with the integrated expression for the attenuation of light intensity by absorption

$$(I = I_0 e^{-nsx})$$

where $s$ is the absorption cross-section (Section 1-3), we can express an inactivation by the relationship

$$N = N_0 e^{-\sigma D}$$

where $N_0$ is the number of active units initially present and $N$ is the number remaining active after a dose $D$. This expression was also introduced in Chapter 1 and it can be derived from the Poisson distribution. The constant $\sigma$ is called the *inactivation cross-section* and it bears no necessary relation to the physical (i.e., geometrical) cross-section of the absorbing species. It is the product of the quantum yield and the absorption cross-section. The expression $\sigma D$ is the average number of hits per sensitive unit. Generally, $\sigma$ will be an order of magnitude smaller than $s$, since there are usually more harmless ways than damaging ones to dissipate absorbed photon energy. The determination of $\sigma$ over a range of wavelengths then constitutes an action spectrum for the particular inactivation effect.

It is evident that this simplest sort of inactivation kinetics involves an exponential loss of active units with increasing photon dose. Thus, the inactivation cross-section may be determined from the slope of the straight line obtained when the logarithm of the surviving fraction, $N/N_0$ is plotted against the dose, $D$. Or $\sigma$ may be obtained from the reciprocal of the dose which gives 37% survival, since $e^{-1} = 0.37$. At this point on the survival curve there is an average of one photon "hit" per sensitive target. An example of a simple "one hit" inactivation curve is given in Figure 6-1. The simplest interpretation of such a curve is that the active unit contains one sensitive site and that one absorbed photon may inactivate the unit. This should not be taken literally to mean that there is only one physical site that is sensitive to photon damage. The sensitive site generally represents the composite of a number of different physical sites of approximately equal sensitivity. The absorption of a photon in any one of these may lead to inactivation of the system. In fact, even if the sensitive sites were of quite different sensitivities, a simple exponential survival curve could be obtained, as long as one hit in such a sensitive site were adequate to cause inactivation.

## 6-3.  BIPHASIC SURVIVAL CURVES

Unfortunately most survival curves do not illustrate the simple exponential one-hit kinetics discussed above. Sometimes the dose-effect curve begins as a simple exponential function but then a break appears and the slope decreases at higher doses. Such a biphasic curve always indicates heterogeneity in the sample, but this can be of different sorts. Thus, one might have a mixture of two types of cells, one type exhibiting greater sensitivity than the other. The more sensitive cells would be more readily inactivated at lower doses thus enriching the mixture with the surviving cells of the resistant type. Eventually almost all of the sensitive cells would be inactivated and the less steep inactivation curve for the remaining resistant cells would prevail. The back extrapolate of the high dose slope to the ordinate then gives an indication of the relative numbers

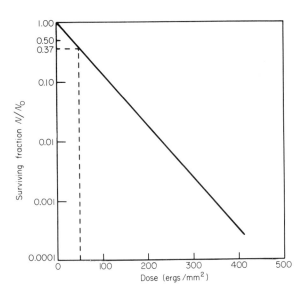

FIGURE 6-1.   An example of simple one-hit inactivation kinetics as expressed by plotting the logarithm of surviving fraction as a function of dose. Such simple exponential kinetics are often expressed in terms of the $e^{-1}$ or 37% survival dose.

of sensitive and resistant cells in the original population (see Figure 6–2).

The heterogeneity may be of three types, broadly classified as genetic, phenotypic, and artifactual. The general method for distinguishing between these classes involves the isolation of survivors after high doses (i.e., in the less steep region of the inactivation curve) and the subsequent determination of a dose-effect curve for this fraction. If the heterogeneity were due to genetic differences, then reirradiation should yield a simple survival curve with a slope characteristic of the more resistant species in the original population. However, sometimes the heterogeneity represents a phenotypic variability within a population of genetically identical systems. Thus, for example, in a growing population, those cells at a particular stage in the division cycle may be more sensitive than those at another stage in the cycle. The survivors from this population after UV and subsequent growth should then exhibit the same variability in response to irradiation as did the original population. Heterogeneity induced by the irradiation itself has not yet been demonstrated although it is theoretically possible. For example, the production of nonlethal events in a system could somehow stimulate the development of resistance to further events (e.g., by inducing the synthesis of repair enzymes). The resistance of the survivors after a given dose would be greater than that of the original unirradiated population. Finally, it is important to consider that the biphasic curve may be due to a heterogeneity in the experimental treatment of the sample. If, for example, the sample were not well stirred during the irradiation some cells might survive by virtue of being in a region which received a lower effective dose than in other regions. Such would be the case for a cell which was temporarily stuck to the bottom of the irradiation vessel where it was shielded by other absorbers from the full intensity of the incident beam. This sort of heterogeneity would not be apparent unless doses of radiation large enough to cause many logarithmic decades of inactivation were used. Furthermore, experimental precautions can be taken to minimize this problem (Section 2-4).

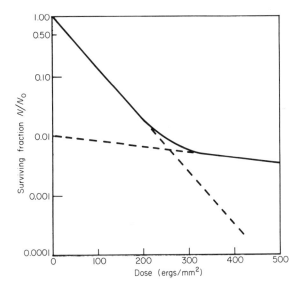

FIGURE 6-2. An example of biphasic or multiple component inactivation kinetics. The less sensitive fraction of the population is determined from the back extrapolate of the high dose curve to the ordinate. This may be subtracted from the observed low dose kinetics to obtain a corrected survival curve for the more sensitive component in the population. In the example shown, this correction would be very small since only 1% of the population is in the less sensitive class.

6-4. MULTIHIT AND MULTITARGET SURVIVAL CURVES

The most common type of survival curve obtained is the one in which the slope increases with increasing dose and eventually ap-

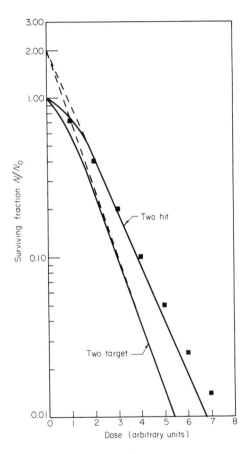

FIGURE 6-3. A comparison of multitarget and multihit inactivation kinetics. The two target curve was calculated on the assumption that the two targets were of equal sensitivity and of the same sensitivity as the one target for the two hit case. The back extrapolate to the ordinate gives the number of targets or number of hits, respectively, for the two cases. If one of the targets is arbitrarily assigned an inactivation cross-section of 74% of that of the other target, then the plotted points (■) more closely approximate the two hit curve. Curve shape alone does not allow the distinction to be made between multitarget and multihit models.

proaches some limiting value. The mathematical treatment of such curves can sometimes be approximated by a multihit or multitarget model for the inactivation process. These are illustrated schematically in Figure 6-3. In the simple one-hit model the expression $e^{-\sigma D}$ gives the probability that the cell will survive. It can be shown that the probability that a cell will receive one hit is expressed by the relationship $\sigma D e^{-\sigma D}$. If two hits must occur to inactivate the cell, then the probability of survival would be the sum of the probability that there were no hits and the probability that there was one hit. Thus, $N/N_0 = e^{-\sigma D} + \sigma D e^{-\sigma D} = e^{-\sigma D}(1 + \sigma D)$ for a two hit process. Or, in general for a $k$ hit process the surviving fraction can be expressed by the relationship

$$N/N_0 = e^{-\sigma D}\left[\sum_{x=0}^{k} \frac{(\sigma D)^x}{x!}\right]$$

However, if we consider the second law of photobiology which states that only one photon is required for a photochemical reaction, then it would seem that multihit processes would have little meaning for the effects of photons on biological systems. A more likely case would be one in which each cell has more than one sensitive component and where a particular number (perhaps all) of these components must be "hit" in order to achieve inactivation of the cell. Suppose the cell had two components of equal inactivation cross-section, $\sigma$. The probability of survival of one of these would be $e^{-\sigma D}$. The probability that it would not survive would be $1 - e^{-\sigma D}$. The probability that two such units would be hit would be the square of this, $(1 - e^{-\sigma D})^2$. Then the probability that the cell would survive (i.e., the probability that both components would not be hit) is $1 - (1 - e^{-\sigma D})^2$. In general, for a multitarget inactivation where $j$ is the number of sensitive components per cell we would have $N/N_0 = 1 - (1 - e^{-\sigma D})^j$. This model, of course, could apply also to the situation in which there were several sensitive components but of different sensitivities, thus:

$$N/N_0 = 1 - (1 - e^{-\sigma D})(1 - e^{-\sigma' D})$$

where $\sigma$ and $\sigma'$, respectively, refer to the several inactivation cross-sections. In the case of a *bona fide* multitarget survival curve the number of targets can be determined from the back extrapolate to the ordinate of the high dose portion of the survival curves as illustrated in Figure 6-3. The value for the component cross-sections can be obtained from the slope of the exponential portion of the curve. The shape of the multitarget curve can be understood intuitively by realizing that after large doses of radiation the probability will be high that almost all of the targets in each cell will have been inactivated. Then the inactivation of the remaining targets (and thus inactivation of the cells) will be a simple exponential function of dose. It is clear from Figure 6-3 that the difference in curve shape for multihit and multitarget inactivations is very subtle indeed. It is impossible in practice to distinguish a multihit curve from a multitarget curve on the basis of the curve shape alone.

In the early development of the target theory for inactivation processes the back extrapolate for survival curves with shoulders was considered to indicate a *multiplicity* of targets. However, it has turned out that many if not most of the apparent multitarget phenomena can be explained by other models. Therefore, it is more appropriate to refer to this back extrapolate as the *extrapolation number*, which may or may not refer to a multiplicity of targets.

In cases in which multiple target kinetics are suspected it is a good idea to obtain independent evidence for the existence of the multiple components. Such a case is illustrated in the "bleaching" of *Euglena gracilis* by UV. Dark-grown cultures of this unicellular green alga are found to contain roughly 30 proplastids. These are self-replicating, DNA-containing, organelles that can develop into chloroplasts when the organism is placed in the light. In fact the UV sensitivity of bleaching provided the first indication that these organelles contained DNA. It has been shown that the development and continued duplication of these proplastids is extremely sensitive to inactivation by UV. The inactivation kinetics for permanent bleaching (i.e., inactivation of all proplastids) follows multiple target kinetics with an extrapolation number of 30 as shown in Figure 6-4.

In contrast to the above case, consider the survival curve obtained for cultures of bacteria in exponential growth compared to that obtained for cultures that have been starved for required amino acids for several generations (Figure 6-5A). The starved culture exhibits a pronounced shoulder in its survival curve. In the example given the extrapolation number is about 240. One might wonder what there are 240 of in a bacterial cell that could be likely candidates for the multitarget inactivation of the cell. Before carrying such speculation very far, it is useful to study the theoretical multi-

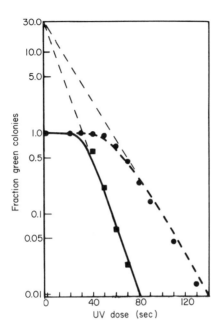

FIGURE 6-4. Kinetics for bleaching of *Euglena gracilis* by UV, an example of a multitarget inactivation curve. The curves shown are theoretical 30 target inactivation kinetics. The points were obtained experimentally. ●, Light-grown cells. ■, Dark-grown cells. [From H. Lyman, H. T. Epstein, and J. A. Schiff, *Biochim. Biophys. Acta* **50**, 301 (1961).]

target curve expected for 240 targets. (Dashed line in Figure 6-5A.) The theoretical curve is much flatter than the experimental curve in the low dose region and, in fact, no appreciable inactivation should have been observed at all in this region if multitarget kinetics were applicable. This shoulder is explained by the presence of an excision-repair system in these bacteria (Section 7-4). A bacterial strain that is deficient in the excision-repair scheme does not present this striking effect following protein synthesis inhibition (Figure 6-5B).

For the sake of completeness, another type of population heterogeneity and its effect upon survival kinetics should be mentioned. Suppose that part of a population of cells contained one sensitive target and thus followed a simple one-hit inactivation, while the remainder of the population contained five sensitive targets. The resultant survival curve for the population might look as illustrated

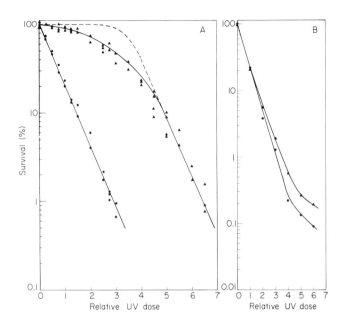

FIGURE 6-5.   The effect of physiological state upon the UV survival kinetics of
*E. coli* strains TAU and $B_{s-1}$: an example in which the multitarget interpretation is
not valid. A. ●, Survival of colony forming ability for an exponentially growing culture
of *E. coli* strain TAU in balanced growth on a chemically defined minimal medium.
Dilutions and platings were done immediately after the irradiation by the pour agar
technique on minimal medium agar plates. ▲ Survival of *E. coli* strain TAU if a 90
minute period of protein synthesis inhibition (by amino acid starvation) immediately
precedes the irradiation. - - - - -, Theoretical 240 target inactivation kinetics for comparison. B. ●, Survival of colony forming ability for an exponentially growing culture
of *E. coli* strain $B_{s-1}$ under conditions equivalent to those for strain TAU above.
▲ , Survival of *E. coli* strain $B_{s-1}$ if irradiation is performed after 2 hours of protein
synthesis inhibition (by addition of 15 μg/ml chloramphenicol). [From P. Hanawalt,
*Photochem. Photobiol.* **5**, 1 (1966).]

in Figure 6-6. The single target component of the population would
be rapidly inactivated before most of the targets in the other component had been hit. Eventually with higher doses the multiple-
target component would begin to be inactivated. An example as
exaggerated as that in Figure 6-6 would be relatively easy to detect
experimentally. However, if the difference in the two components
were less pronounced it might be very difficult to distinguish this
kind of heterogeneity from a single exponential inactivation curve.

If one had a population of cells in which some had one sensitive nucleus while others had two sensitive nuclei, then the above considerations might apply. The inherent inaccuracy of the colony count assay for viability might prevent the detection of a slight displacement of the exponential portion of the curve at higher doses. The data would in practice be approximated by a simple exponential curve and the averaged value for the inactivation cross-section would be slightly too small. For exponentially growing cultures of bacteria it has been determined by cytological and other methods that some of the cells do contain more than one nuclear region. Yet, as evident in Figure 6-5A and B, simple exponential survival curves can be obtained with growing cultures of some strains. Also, in

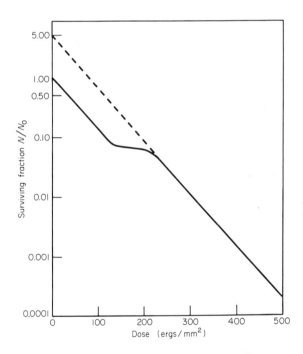

FIGURE 6-6. Kinetics for inactivation of a heterogeneous population containing components of different multiplicities (1 target and 5 targets) but with the same target sensitivity. Ten percent of the population is in the multitarget class. Such curves can easily be generated artificially (e.g., by mixing cultures as represented by the two curves in Figure 6-5A).

Figure 6-5B, an exponential survival curve is obtained following a period of protein synthesis inhibition leading to a physiological state in which all cells have two nuclear regions. The evident conclusion is that the presence of two nuclei in bacteria does not necessarily result in the two target situation for UV inactivation. On the other hand, a clear distinction between the X-ray inactivation curves for haploid and diploid yeast has been found, corresponding to single target and two target kinetics, respectively.

## 6-5. BACTERIAL VIABILITY

It may be concluded from the above that very few survival curves can be used as a direct indication of the damage inflicted, even if one is equipped with a complete catalogue of the photochemical effects and their relative yields. In Figure 6-7 are compared the survival curves for the three closely related bacterial strains, *E. coli* B/r, *E. coli* B, and *E. coli* $B_{s-1}$. Even though the initial photochemistry is identical (i.e., same number of pyrimidine dimers produced for same UV dose) the biological response is seen to be quite drastically different. The range of UV sensitivities represented by these three strains is quite typical of that found among bacterial strains. *Micrococcus radiodurans*, however, has been found to have an amazing resistance to UV, in comparison with all other bacteria tested. The survival curve for this strain is shown in Figure 6-8.

The explanation for its radiation resistance is the presence of an extremely efficient system for repairing pyrimidine dimers. The killing is probably due to other types of photochemical damage as discussed in Section 4-9. This again emphasizes the important point that the relative biological effectiveness of different photoproducts may vary depending upon the response of the organism to them.

Some biological systems can respond to adverse growth conditions by entering a special physiological state in which metabolism is all but shut down and in which the resistance to a variety of deleterious effects is markedly increased. In bacteria this process involves the formation of spores and in protozoans the analogous state is referred to as a cyst. Spores exhibit striking resistance to heat inactivation and also to inactivation from radiation of all sorts. Little or no cyclobutane-type pyrimidine dimers are formed in UV-irradiated bacterial spores but a new photoproduct of thymine is

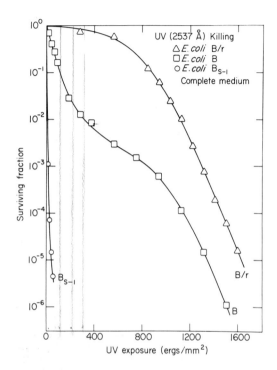

FIGURE 6-7. UV survival curves for several closely related strains of *E. coli*. [From R. H. Haynes, *Photochem. Photobiol.* 3, 429 (1964).]

found. Thus, much of the pronounced resistance in spores is probably due to a different class of photochemical products rather than a different manner of dealing with certain photoproducts (Section 4-6).

Sometimes it is possible to vary the initial photochemistry by altering the physical conditions in the environment. This can change the relative preponderance of different biologically effective photoproducts and thus affect the sensitivity. This is well illustrated by the effect of temperature on the inactivation of bacteria previously discussed in Section 4-9.

## 6-6. MACROMOLECULAR SYNTHESIS

The measurement of viability, or more specifically the determination of the ability to form colonies on a nutrient agar plate, is one

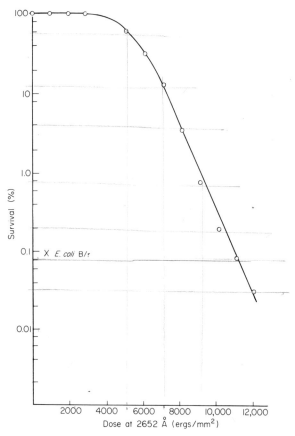

FIGURE 6-8. The UV survival of *Micrococcus radiodurans* at 2652 Å. The "X" indicates the dose (2537 Å) which inactivated *E. coli* B/r to 0.1% survival in the experiment shown in Figure 6-7. [Adapted from J. K. Setlow and D. E. Duggan, *Biochim. Biophys. Acta* 87, 664 (1964).]

of the most convenient methods of assay for the sensitivity of bacteria to UV, and this is perhaps one reason that it is used so extensively. However, it must be borne in mind that there are many other parameters that might equally as well be considered in a study of UV sensitivity. Just as the assay of colony count alone does not provide a complete picture of cell growth, so also it does not provide a complete picture of cell inactivation. Viability is one of the more sensitive parameters to UV inactivation, in part because of the multitude of sensitive phenomena essential to the duplication of a viable cell. It is not, however, the *most* sensitive parameter. In some strains of bacteria, of which *E. coli* B is the prime example, the ability to divide within a few hours after irradiation is much more sensitive than is colony-formation (usually measured 24 hours later). This is not easily picked up by the colony-count assay because many (and in some cases, most) of the division-delayed cells will eventually resume normal division and thus will give rise to normal colonies. The observation of the cells under a microscope an hour after the irradiation, however, will make it quite clear that many are becoming longer without dividing. This phenomenon of "filament formation" will be considered further in Section 7-5.

Probably more relevant to an understanding of mechanisms of biological inactivation are the specific effects of UV upon macromolecular synthesis in cells. Examples of the effect of UV upon DNA synthesis in certain radiation resistant and sensitive strains of bacteria are shown in Figure 6-9. The assay for DNA synthesis used in these studies was the incorporation of tritium-labeled thymidine. The radioactive thymidine was added to the culture of growing cells immediately after irradiation and the extent of incorporation of the radioactive label into macromolecular DNA was then determined after various periods of growth. Data of the sort represented in Figure 6-9 clearly show that the effect of increasing doses of radiation on bacterial cultures is to reduce the amount of radioactivity incorporated in a given period of time. The detailed interpretation of the apparent kinetics of isotope incorporation is more difficult.

The results in Figure 6-9A have been interpreted as indicating that DNA synthesis proceeds up to a dimer (*E. coli* $B_{s-1}$ cannot excise dimers) and then synthesis stops completely and permanently. The strains shown in Figure 6-9B, C, and D can all excise dimers (although *E. coli* $B_{s-3}$ does so very inefficiently) and the shapes of their curves indicate that DNA synthesis is inhibited but that it resumes at near normal rate at a time which is dependent upon the dose of UV. The dose required for a given delay in synthesis is markedly different for the three strains, and may give a measure of the relative efficiency for repair in the different strains. From this type of graph, delays in synthesis much longer than a generation time have been inferred.

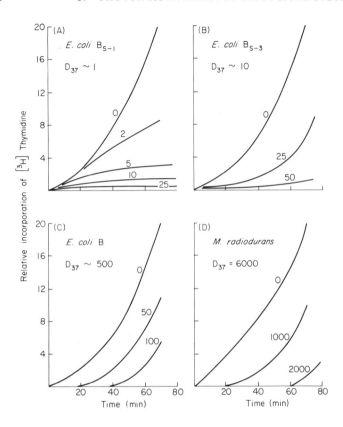

FIGURE 6-9. The effect of ultraviolet radiation (265 nm) on subsequent DNA synthesis in irradiated cultures. The numbers next to the curves and the values of $D_{37}$ represent doses in ergs/mm$^2$. *E. coli* $B_{s-1}$ is *hcr*$^-$ and excision minus; *E. coli* $B_{s-3}$ is *hcr*$^-$ but has a low level of excision. *E. coli* B/r and *Micrococcus radiodurans* both excise dimers very efficiently. [From R. B. Setlow, *in* "Regulation of Nucleic Acid and Protein Biosynthesis" (V. V. Koningsberger and L. Bosch, eds.), p. 51. Elsevier, Amsterdam, 1967.]

Linear plots of counts per minute (cpm) of incorporated radioactivity vs. time can lead to incorrect conclusions concerning DNA synthesis kinetics. Curve shapes similar to those represented in Figure 6-9B, C, and D can be closely approximated by using *unirradiated* cultures starting at different cell densities (Figure 6-10). The apparent lags in DNA synthesis shown in the linear plot in

Figure 6-10 are eliminated when the data are plotted on a semi-log scale.

The data in Figure 6-10 indicate some of the experimental and graphical variables that can affect the *apparent* kinetics (but not the *true* kinetics) of DNA synthesis in an *unirradiated* bacterial culture. The problem of understanding DNA synthesis kinetics in an irradiated culture of bacteria, however, is even more complex because of the UV induced heterogeneity in the population of cells. One can postulate the presence of three major classes of cells after irradia-

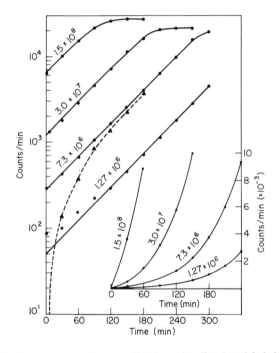

FIGURE 6-10. The uptake of thymine-($^{14}$C) by unirradiated prelabeled cultures of *E. coli* B/r, T$^-$, having different initial numbers of colony forming units (CFU). In the linear plots (corrected to zero counts at zero time) shown in the lower right-hand corner the rates of thymine uptake *appear* different for cultures containing different initial numbers of CFU. In contrast, the semi-log plot shows correctly that the rates are independent of cell number. The dotted curve (log scale) indicates the type of result obtained if one uses a semi-log plot without using prelabeled cells (mimicked here by subtraction of the zero time counts from all subsequent samples). [From K. C. Smith and M. E. O'Leary, *Biochim. Biophys. Acta* 169, 430 (1968).]

tion: (1) colony formers synthesizing DNA; (2) noncolony formers synthesizing DNA for variable amounts of time; and (3) noncolony formers not synthesizing DNA. Classes (1) and (2) can be subdivided according to their rate of DNA synthesis (delayed, normal, or reduced). Therefore, the interpretation of DNA synthesis kinetics and its relation to growth kinetics would appear to be complicated at best, even if optimal experimental design and graphing procedures are used.

A new protocol has been formulated to simplify the interpretation of isotope incorporation studies: (1) use mutants requiring the labeled precursor (e.g., thymine), prelabel the cells and keep the radioactive precursor present after irradiation (the incorporated radioactivity then becomes a direct measure of the amount of DNA and is equivalent to the direct chemical determination of DNA); (2) plot the data as log counts per minute vs. time (DNA synthesis is an exponential function in bacterial cultures); and (3) determine the kinetics of cell growth at the same time aliquots are taken for the determination of incorporated radioactivity (one may then hope to deduce the biological relevance of the observed synthesis).

Using this protocol, the data for *E. coli* $B_{s-1}$, $T^-$ (Figure 6-11) make it clear that dimers do not stop DNA synthesis completely and permanently. Synthesis proceeds at essentially the normal rate for up to about 15 minutes (depending upon the dose of UV) and then the rate falls off. This is consistent with the results in Figure 6-9A showing an apparent cessation of DNA synthesis. However, during further incubation DNA synthesis resumes and increases to rates which are inversely correlated with dose (Figure 6-11).

The fact that dimers do not constitute absolute blocks to DNA synthesis in excision-minus strains has also been demonstrated by physical-chemical methods. Using an excision-minus strain of *E. coli* K-12, Rupp and Howard-Flanders showed that the DNA synthesized for the first 10 minutes after UV irradiation consists of smaller single-strand segments than those produced under similar conditions in unirradiated cells. (The comparison was made by zone sedimentation of the isolated DNA in alkaline sucrose gradients in an ultracentrifuge). Upon further incubation of the cells (in nonradioactive media), however, the low molecular weight DNA labeled during the 10 minute pulse attains essentially the same size distribution as in the control. It has been proposed that a break is left in the

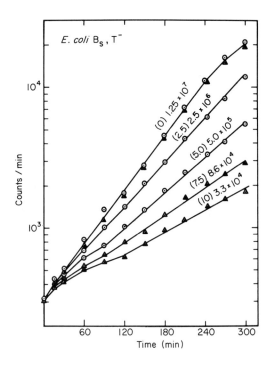

FIGURE 6-11. The uptake of thymine-$^{14}$C by UV-irradiated (2537 Å) prelabeled cultures of *E. coli* B$_{s-1}$ T$^-$. The numbers in parenthesis indicate the radiation dose in ergs/mm$^2$. The adjacent numbers indicate the number of survivors as determined on salts-glucose agar plates supplemented with thymine. The results are plotted as counts per 0.05 ml of culture. [From K. C. Smith, unpublished observations (1968).]

daughter strand as replication proceeds past a dimer (in the parent strand) and that these gaps are subsequently repaired by a rejoining event. Ganesan and Smith have shown that certain excision-minus strains of *E. coli* K-12 exhibit a large recovery in viability after UV irradiation when they are grown in minimal medium instead of complex medium. This recovery in minimal medium is not observed for excision-minus strains that are also recombination deficient. The molecular basis of this mode of recovery is suggested by the observations of Rupp and Howard-Flanders, and by the fact that K. C. Smith *et al.* have shown that mutants which are both excision-minus and recombination deficient do not join together the small pieces of DNA synthesized after UV-irradiation.

Again, using the new protocol for measuring DNA synthesis kinetics, the data shown in Figure 6-12 were gathered on *E. coli* B/r. Contrary to previously published studies using radioactive precursors, the results in Figure 6-12 demonstrate that DNA synthesis is not stopped for times longer than one generation time. Doudney and Young reported some years ago that DNA synthesis was not completely inhibited by UV irradiation for longer than one generation time in a radiation resistant strain of *E. coli*. They assayed for DNA by a colorimetric method.

It would appear that some significance should be attached to the observation that UV irradiation does not stop DNA synthesis in excision-minus cells (*E. coli* B$_{s-1}$) but does so in excision positive cells (*E. coli* B/r) for periods shorter than one generation time. In attempting to understand these observations it must be borne in

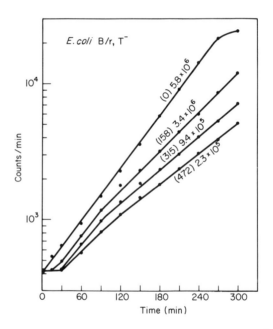

FIGURE 6-12. The uptake of thymine-$^{14}$C by UV-irradiated (2537 Å) prelabeled cultures of *E. coli* B/r, T$^-$. The explanation of numbers and symbols is given in Figure 6-11. [From K. C. Smith and M. E. O'Leary, *Biochim. Biophys. Acta* 169, 430 (1968).]

mind that the dose to kill 63% of the population (final straight portion of the survival curve) is about 75 times greater for *E. coli* B/r than for B$_{s-1}$. Furthermore, *E. coli* B/r shows a UV-induced inhibition of cellular division which can persist for up to about 3 generation times before division resumes at essentially the normal rate. *E. coli* B$_{s-1}$, however, shows no lag in cellular division after irradiation. If cell division is dependent upon DNA synthesis then why should cell division continue to be inhibited (in *E. coli* B/r) after DNA synthesis resumes? Perhaps it takes longer to repair the lesions leading to cell division inhibition than those leading to DNA synthesis inhibition.

The new methodology for determining macromolecular synthesis kinetics in irradiated cells (including simultaneous determination of growth kinetics) particularly if applied to specific mutants sensitive to UV radiation (such as excision-minus and recombination deficient strains) should provide further insights into the intricacies of DNA synthesis and its control.

In spite of the difficulties in the detailed interpretation of kinetic data for macromolecular synthesis in bacterial cultures following irradiation, the application of such methods has yielded information on the relative sensitivities of different macromolecular syntheses to irradiation. The doses of UV (2537 Å) required to inhibit RNA and protein synthesis are higher than those required to inhibit DNA synthesis in bacteria. It might be expected that the same sort of block that would inhibit the passage of the DNA polymerase on the DNA template would similarly inhibit the passage of the RNA polymerase on that template. However, transcription could continue at a number of sites in the DNA that did not contain such blocks, while the sequential progress of the few normal DNA replication points would soon be affected.

Action spectra for the effects of photons upon macromolecular synthesis in bacteria have generally implicated nucleic acid as the sensitive target, and this is of course, quite expected and consistent with the role of nucleic acid as the information carrying species. In the very UV-resistant bacterium, *Micrococcus radiodurans*, however, the action spectra for killing and for inhibition of DNA synthesis show a high involvement of protein as well as nucleic acid.

Although DNA is the target most likely to be responsible for the permanent loss of viability in cells it must be remembered that the various classes of RNA may be similarly sensitive to loss in function when damaged by photons. The transfer RNA species, for example, contains a number of specific recognition sites: the anticodon nucleotide triplet recognizes the codon triplet on a messenger RNA, another site must specifically recognize an appropriate amino acid activating enzyme, and still other sites may be involved in the binding and orientation of the transfer RNA in the ribosome. It has been shown that the transfer RNA is inactivated by UV and that the uracil and cytosine residues are the sensitive targets. The inactivation of the amino acid acceptor function of several transfer RNA's from *E. coli* has been found to parallel closely the UV-induced conformation changes in the molecule as measured by optical rotatory dispersion. It would appear that the UV-induced conformational changes which affect recognition of the amino acid activating enzyme are more sensitive to inactivation than is the anticodon. In a similar manner one might speculate that a UV-induced conformational alteration in ribosomal RNA could result in the failure of the ribosome to properly align the incoming loaded transfer RNA's. However, the action spectrum for the inactivation of ribosomes to synthesize polyphenylalanine (using poly-uridylic acid as the messenger) shows a peak both at 2600 Å and 2800 Å suggesting the involvement of protein as well as ribonucleic acid in inactivation. The ribosome might well be further unveiled through the use of the UV photobiological probe, and this may be a fruitful area for investigation.

6-7. BACTERIOPHAGE

The virus is an attractive biological entity for photobiological investigation. Since it can generally be purified (indeed crystallized!), very accurate determinations of photochemical events can be made without the complications of metabolic activity or repair processes. The treated virus can then be turned loose on its appropriate host cell and the resultant biological phenomena can be assayed. In some respects one achieves similar advantages in working with isolated transforming DNA in photobiological studies. However, the transforming system is largely dependent upon recombination events for the biological assay, whereas the virus can be examined with respect to many biological phenomena.

One might expect that a relatively uncomplicated system like a bacterial virus would exhibit a simple exponential loss in infectivity as a function of UV dose. But this is often not the case, and for a variety of reasons. Some of these are trivial while others provide important information on the structure and mode of replication of the phage genome.

Phage T4 is inactivated by UV with nearly exponential kinetics, but there is sometimes a slight shoulder on the survival curve. A closer examination of this phenomena revealed that the extent of the shoulder was a function of wavelength. (This alone should be sufficient to cause one to discard the possibility of multiple targets.) It was found further that more extensive purification of the phage resulted in the complete elimination of the shoulder. It has been suggested that the UV activated a small fraction of the population of phage that had been rendered inactive by some inpurity in the preparation. This effect is rather trivial in terms of the biology of bacteriophage.

A far more important phenomenon was discovered by Luria in the course of photobiological studies on the T4 phage system. He observed that the plaque forming ability of the phage seemed to become relatively constant at very high doses of UV and he subsequently showed that this was the result of the infection of a bacterium by two or more *inactive* phage. The inactive phage could somehow cooperate to produce one active phage which eventually gave rise to an observable plaque. This phenomenon, termed *multiplicity reactivation,* is also discussed in Section 7-2, as an example of "fortuitous" recovery from damage to the genome since there is no specific recognition of the damage in the inactive phage. The near random process of genetic recombination among the inactive particles can result in the assembly of an active particle. Actually, the radiation assists in this process by increasing the frequency of genetic recombination.

One might then think of the inactivation of phage in terms of an expression of the sort:

$$N/N_0 = e^{-(F_1\sigma + F_2\sigma + \ldots F_n\sigma)D}$$

where the $F$'s are the fractions of the total phage lethal hits due to inactivation of the specific targets 1 to $n$. With this model in mind early multiplicity reactivation experiments were designed to determine the number $n$. However, there was the very basic difficulty that the multiplicity of infection can not be controlled in practice and so it represents only an average. As the average multiplicity is increased, the UV sensitivity for plaque formation is decreased. An equation was derived by Luria and Dulbecco to describe theoretically the probability of phage production in a multiply-infected cell. It was predicted that the slope of the survival curve as a function of phage lethal hits should be constant for large doses, independent of the multiplicity, and equal to the slope found for single infecting phage particles. The back extrapolate to the ordinate should yield the same value of $n$ (the number of sensitive sites), as the multiplicity of infection is varied. In support of the hypothesis were the data obtained for the UV-irradiated Vi-phage II of *Salmonella*. But for phage T2 the slope of the survival curves at high multiplicities of infection was found to be significantly less than that for the inactivation of free phage (Figure 6-13). An explanation was suggested in which it was assumed that about 60% of the sensitive sites in T2 could be reactivated with an efficiency so high as to have no effect on the curves. The other 40% was assumed to be composed of three vulnerable centers. These centers were postulated to be those involved with functions that were essential for the multiplication and/or recombination of the vegetative phage. If these centers were hit the damaged phage would not be able to undergo the multiplicity reactivation essential for survival. Sixty percent of the sensitive sites presumably involved phage functions that were not needed until recombination had occurred.

The assistance of one disabled phage by another need not involve genetic recombination. In a multiply-infected cell an enzyme produced by an otherwise inactive phage may help another phage that is inactive because of its inability to produce that enzyme. Thus, an irradiated phage may still perform many of the functions of a viable phage even though it appears "dead" by the plaque forming assay. For example, a UV dose that reduces infectivity in phage T2 to $10^{-6}$ survival only reduces the ability of the phage to kill the host to about 35%. One cannot generalize from this example, however, since for the phages T1, T3, and T7 host-killing is nearly as sensitive as plaque formation.

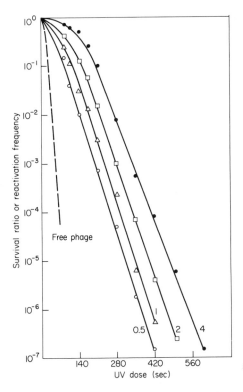

FIGURE 6-13. Multiplicity reactivation of phage T2 following UV irradiation. The numbers by the curves indicate the average multiplicity of infection in the bacterial population. The dashed line indicates the survival curve for free phage and the solid lines indicate the survival of phage in multiply-infected bacteria at the multiplicities indicated. [From S. E. Luria, *in* "Radiation Biology" (A. Hollaender, ed.), Vol 2, p. 333, McGraw-Hill, New York, 1955, as modified from R. Dulbecco, *J. Bacteriol.* **63**, 199 (1952).]

A particularly useful way in which the UV photon probe can be used is in the so-called "Luria-Latarjet" experiment in which the complex of phage and bacteria is irradiated at different times during the vegetative cycle following infection. The UV sensitivity of phage T4 within infected bacteria is initially like that of free phage irradiated before infection. However, upon irradiation 5 minutes after infection there is a very dramatic decrease in phage sensitivity. A number of possible explanations for this phenomenon have been considered; (1) replication of the phage DNA may have permitted multiplicity reactivation; (2) a repair system may have been activated by the infecting phage; (3) the target size of different essential genes may decrease as these genes are progressively expressed and become dispensable; or (4) the physical structure of the DNA itself may undergo changes during development that affect its sensitivity. The detailed explanation of the phenomenon has been hindered by the fact that the plaque forming assay is a delayed response and is sensitive to several different causal events (e.g., blockage of DNA replication or the expression of vital genes). In such a situation it is useful to find an assay that more directly relates to the molecular event that is responsible. In one such study the sensitivity of synthesis of a specific protein, phage lysozyme, was examined in the T4 system. Lysozyme is a so-called "late" enzyme and it is synthesized about 10 minutes after phage infection. As shown in Figure 6-14 the sensitivity of lysozyme synthesis has decreased considerably between the time of phage infection and 5.5 minutes later. Closer examination has revealed an abrupt change in sensitivity for lysozyme synthesis between 4 and 6 minutes after infection. Thus, lysozyme production seems to become more UV resistant at the same time in development as does plaque-forming ability. It has been suggested that the highly sensitive tar-

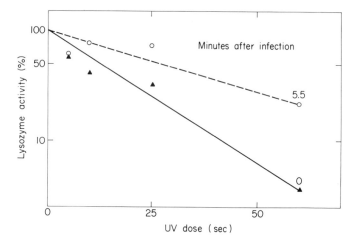

FIGURE 6-14.    UV inhibition of T4 lysozyme synthesis in *E. coli* B after infection with the phage. [From D. B. Fisher and A. B. Pardee, *Virology* 34, 91–96 (1968).]

get is not the gene for lysozyme per se but rather some region of the genome that regulates the synthesis of the "late" enzymes.

## 6-8. INDUCTION OF PROPHAGE

In our discussion thus far we have considered only the virulent bacteriophage. These phage undergo the lytic cycle involving multiplication of the viral genome and eventual lysis (bursting) of the cell with the concommitant release of several hundred progeny phage. There is another class of bacteriophage, the so-called temperate phage. They may initiate the lytic process or they may undergo lysogeny, a phenomenon that entails the physical insertion of the phage genome into the bacterial chromosome. In lysogeny the "early functions" in phage multiplication and development do not occur and instead the integrated phage genome (called the prophage) is replicated as part of the bacterial chromosome. Most prophage are integrated into the bacterial chromosome at a specific site, presumably by recombinational events that require pairing of homologous regions of phage and bacterial DNA. The decision as to whether the phage undergoes the lytic or lysogenic response depends upon a delicate equilibrium between the phage directed synthesis of proteins required for "early functions" and the phage directed synthesis of a special protein called the *repressor*. The repressor specifically binds to a unique region in the phage genome and thereby prevents the transcription of the genes required for viral multiplication. The lysogenized bacterium can proceed to grow for many generations, essentially as an uninfected cell. (The prophage may make its presence known by certain genetic phenomena — e.g., by preventing superinfection of the bacterium by another of the same kind of phage.) Infrequently the lysogenic bacterium will release its prophage spontaneously and the lytic response will ensue. For several years Andre Lwoff, the discoverer of lysogeny, tried unsuccessfully to induce lysogenic bacteria to enter the lytic cycle. After trying every chemical on his shelf he looked at the effect of a low UV dose and was amazed to discover that essentially all of the bacteria were producing phage. The action spectrum for induction was found to implicate nucleic acid, and the photoreactivability of induction specifically indicated DNA as the target.

If direct inactivation of the repressor by UV were the mechanism, then the action spectrum should have reflected the protein nature of the sensitive unit. Recent studies have suggested an indirect mechanism for the inactivation of the repressor. It is instructive that the agents that stimulate induction are almost without exception those agents that damage DNA and lead to repair replication. (Heat induction is an exception and involves direct inactivation of the repressor protein.) The evidence suggests that the inducing event leads to the inactivation of the repressor, rather than the *de novo* synthesis of an "inducer." How does the UV photochemistry in some nucleic acid (probably DNA) lead to the inactivation of a specific protein, the repressor? This currently stands as an unsolved problem in UV photobiology.

The phenomenon of induction is also important from the standpoint of the survival of the host cell. Prophage induction is clearly an indirect means by which a living cell may be inactivated by UV. In some cases this inactivation need not be marked by the subsequent appearance of infective particles. The phage may be defective, such that they multiply and kill the cell and perhaps even cause lysis but they may not be capable of further infection. This possibility leaves the experimenter with the uncomfortable feeling that he can never be certain that the UV inactivation of a cellular system does not involve prophage induction.

6-9. PLANT VIRUSES

Single hit kinetics are observed for the UV inactivation of infectivity in tobacco mosaic virus (TMV). It is curious, however, that a marked concentration dependence for the quantum yield is seen. The quantum yield is decreased roughly by a factor of four as the concentration of irradiated TMV is reduced 4-fold to 10 $\mu$g/ml. In the 40 $\mu$g/ml to 100 $\mu$g/ml range no difference in quantum yield is found. It is not clear whether this concentration effect is due to aggregation or to some mutual photosensitization phenomenon. It is therefore best to determine quantum yields at several different concentrations and to extrapolate to zero concentration to obtain an intrinsic quantum yield for the virus particle. Other systems that tend to aggregate may be expected to show such effects of concentration on UV sensitivity.

The action spectrum for the loss of infectivity of isolated RNA from TMV looks like the absorption spectrum for the RNA. However, the action spectrum for the loss of infectivity of the intact

TMV particle shows an additional protein component. The protein sheath of the virus must be stripped off early in the infective process. Either the inactivation of the RNA genome or the UV induced denaturation of the protein coat may prevent the completion of this process. In spite of the additional protein component in the UV inactivation of the intact virus particle, the quantum yield is actually less than that for the RNA. Also, there appears to be no transfer of energy absorbed in the protein to the RNA moiety. Thus, the presence of the protein sheath does help to protect the virus RNA from UV inactivation. At least part of this protective effect may be due to the tight folding of the RNA within the coat, so that formation of certain photoproducts (e.g., dimers) are less sterochemically feasible while others (e.g., hydration products) may be less likely because of the exclusion of water. The UV sensitivity of plant viruses is generally not directly correlated with their nucleic acid content. This is not too surprising in view of the importance of the protein coat to the sensitivity. There exist several strains of TMV which differ markedly (i.e., 6-fold) in UV sensitivity but which yield infectious RNA of identical sensitivities.

It is possible to study the UV sensitivity of TMV during the infective process just as has been done with phage-infected bacterial cells. Dosimetry becomes a more serious problem in the study of plant virus infection, because of the UV absorption by the leaves. A simple exponential UV survival curve is obtained for the virus in infected leaves for about 5 hours after inoculation. Then the curve assumes a multitarget shape as replication gives rise to additional infectious particles. UV is a useful probe for determining the time course of appearance of new infective units. During the initial 5-hour period after inoculation the sensitivity is at first the same as for the free virus, then the resistance increases until the progeny begin to appear. If plants are infected with free RNA very little lag is seen before the increase in UV resistance occurs.

6-10. ANIMAL VIRUSES

Much less photobiology has been done with the animal viruses than with plant virus and bacteriophage systems. The various biological activities of influenza virus and their UV sensitivities are illustrated in Figure 6-15. The action spectra for some of these parameters is quite different: (Figure 6-16) infectivity, as expected,

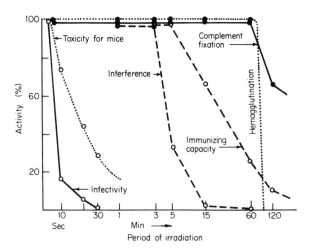

FIGURE 6-15.   Some of the UV-sensitive parameters that may be studied in the influenza virus and their comparative survival curves. Irradiation at 2537 Å. Composite curves from several sources.

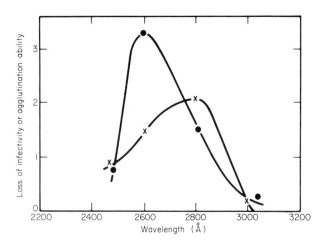

FIGURE 6-16.   Action spectra for the inactivation of infectivity (●) or of hemagglutination (x) in influenza virus. [I. Tamm and D. Fluke, *J. Bacteriol.* **59**, 449–461 (1950), as modified by A. D. McLaren and D. Shugar, "Photochemistry of Proteins and Nucleic Acids," p. 292. Pergamon Press, Oxford, 1964.]

yields principally a nucleic acid action spectrum while hemagglutination gives rise to a protein-like action spectrum. The latter is expected since hemagglutination is a surface property of the virus. Surface properties of the virus are also involved in its action as an antigen and this immediately suggests the possibility of using UV for the production of vaccines. It should be noted that infectivity is much more sensitive to UV than is immunizing capacity (Figure 6-15).

## 6-11. EUCARYOTIC CELLS

Photobiology becomes considerably more complex when we are dealing with large eucaryotic cells or cells in a multicellular organism. In such cells the photochemistry in molecules other than the nucleic acids of the nucleus may become important in inactivation effects. This is so for a number of reasons. Eucaryotic systems often have more redundancy of genetic information so that the inactivation of a particular gene is less likely to be consequential. They are also more highly differentiated (in a multicellular system) so that there are fewer functions that are critical for survival of a given cell. Finally, the nuclei in such cells are shielded by UV absorbing material in the surrounding cytoplasm.

If one is used to microbial systems one is impressed by the difference in time scale in eucaryotic systems. Whereas a given dose of UV may inhibit DNA synthesis for about 40 minutes in *E. coli* the same dose may inhibit DNA synthesis for 40 hours in strain L mouse cells in tissue culture. (The technique of tissue culture helps to surmount the problems of organism complexity and of their contribution to photobiological effects at the cellular level.) Strain L cells normally grow with about a 20 hour generation period in tissue culture. DNA synthesis occupies only 25 to 30% of their growth cycle. Cultures of these cells can be induced to divide synchronously. Thus, it is possible to look at UV sensitivity during the generation cycle at times when DNA synthesis is not occurring. This is an advantage over the bacterial systems, in which DNA synthesis is essentially continuous throughout the generation cycle. As with bacterial systems it has been found that division ability is one of the most sensitive parameters to inactivation by UV. However, the results are very dependent upon the culture conditions: for cells having a relatively high DNA content division appears to be less sensitive than DNA synthesis inhibition. Cells with low DNA content have been found to be relatively more sensitive to UV inhibition of division.

The ultraviolet sensitivity of eucaryotic cells has been found to vary during the mitotic cycle. The developing egg of *Ascaris* is more sensitive to UV than the unfertilized egg and a maximum sensitivity is attained about two hours after fertilization. However, after the first cleavage the sensitivity returned to that of the unfertilized egg. Similar studies on the sea urchin egg have also indicated an enhanced sensitivity after fertilization and prior to the first cleavage. The eggs became more resistant during the metaphase period and remained resistant until division. Neuroblast cells from the grasshopper are sensitive to the retarding of the mitotic cycle by UV if they are irradiated during interphase or prophase. Since DNA synthesis is occurring during the interphase period this result is perhaps not surprising—if indeed DNA is the principal target for this effect. It is clear that there is much to be done to bring our understanding of UV-inactivation effects in such systems to the degree of sophistication attained with viruses and bacteria. Nevertheless, the photon probe may prove to be an important tool for unravelling the mysteries of these complicated cells.

One specific approach that can be used successfully with large eucaryotic cells involves the selective irradiation of parts of a cell. This can be accomplished by the use of a microbeam apparatus that is essentially a microscope with an additional optical system for focusing the beam from a UV source upon a selected portion of the sample. UV microbeam irradiation has been used to demonstrate that irradiating the nucleus is more deleterious to the survival of the cell than is irradiating the cytoplasm. If just one of the two nucleoli in the grasshopper neuroblast is irradiated in early prophase there is a permanent cessation of mitosis. Microbeams have been directed at various components of the mitotic apparatus to show, for example, that a chromosome loses its ability for directed movement if its kinetochore is irradiated. The irradiation of the nucleus of *Euglena gracilis* inhibited cellular division but did not have any effect upon development of chloroplasts, while selective bleaching of chloroplasts could be accomplished with no effect upon cell division. The use of the UV photon probe for the study of eucaryotic cell systems has not achieved the same degree of sophistication that has been attained with the viruses and bacterial systems. In principle UV can be used for obtaining detailed information on the functional organization of these complicated systems and this remains an open and relatively unexplored field of research.

## GENERAL REFERENCES

D. E. Lea, "Actions of Radiations on Living Cells," Cambridge Univ. Press, London and New York, 1955. (The basis of target theory.)

S. E. Luria, Radiation and viruses. *In* "Radiation Biology" (A. Hollaender, ed.), Vol. 2, p. 333. McGraw-Hill, New York, 1955.

C. P. Swanson and L. J. Stadler, The effect of ultraviolet radiation on the genes and chromosomes of higher organisms. *In* "Radiation Biology (A. Hollaender, ed.), Vol. 2, p. 240. McGraw-Hill, New York, 1955.

M. R. Zelle and A. Hollaender, Effects of radiation on bacteria. *In* "Radiation Biology" (A. Hollaender, ed.), Vol. 2, p. 365. McGraw-Hill, New York, 1955.

R. E. Zirkle, Partial-cell irradiation. *Advan. Biol. Med. Phys.* **5**, 103 (1957).

A. C. Giese, Studies on ultraviolet radiation action upon animal cells. *In* "Photophysiology" (A. C. Giese, ed.), Vol. 2, Chapter 17, p. 203. Academic Press, New York, 1964.

H. H. Seliger and W. D. McElroy, "Light: Physical and Biological Action." Academic Press, New York, 1965.

# 7

# Recovery from Photochemical Damage

7-1. INTRODUCTION

Very early in the process of evolution mechanisms must have arisen to protect cells and to facilitate recovery from the damaging effects of photons. In this chapter we will outline the various devices by which organisms are able to minimize the predominantly deleterious effects of photons that threaten survival. Although we will be concerned with repair mechanisms primarily from the point of view of recovery from photochemical damage, the more general implications of these mechanisms should be borne in mind and these will be emphasized where suitable comparisons can be made.

There is a very basic problem in the detection of a repair mechanism, namely that there is no certain method for determining that such a mechanism is operating unless there is some way to turn it off or at least to reduce its effectiveness. The search for recovery mechanisms involves changing the conditions of treatment and environment so that the efficiencies of possible recovery mechanisms may be altered. The most powerful methods involve, in addition, the use of UV-sensitive mutants that might be deficient in one or more steps in a repair process. This approach provided the first proof of the presence of a dark-repair mechanism in bacteria.

Before considering specific recovery mechanisms we should outline the possible modes of molecular recovery in general terms. Three modes which have been documented are the following:

(a) The damaged molecule or part of a molecule may be restored to its functional state *in situ*. This may be accomplished by the activity of some enzymatic mechanism or it may simply result from the "decay" of the damage to an innocuous form.

(b) The damaged unit may be removed from the molecule or system that contains it and replaced with an undamaged unit to restore normal function.

(c) The damage may remain unrepaired in the system, but for one reason or another the system may be able to bypass or ignore the damage.

It is not sufficient to consider recovery merely in terms of the repair mechanism itself in order to relate it to the ultimate survival of function in the organism. In addition, the conditions under which recovery is taking place must be specified. It has become increasingly clear that conditions favorable to survival will require that any molecular repair process be completed within some narrowly defined period of time in order to be effective. Thus it may be too late to repair the damage to a biological molecule *after* that damaged molecule has attempted the performance of its required activity in the cell. It is also important to appreciate that the relative contributions of different photoproducts to the biological effect may change as a function of the extent and efficiency of repair, since different photoproducts may be repaired with different efficiencies and some may not even be repairable.

The first indication of possible recovery phenomena in connection with photochemical damage came from the studies of Hollaender and Claus who found that higher survival levels of UV-irradiated fungal spores could be obtained if they were allowed to remain in water or salt solution for a period of time before plating on nutrient agar. More than ten years later Roberts and Aldous extended these observations by showing that the shapes of UV survival curves for *E. coli* B could be changed quite drastically by simply varying the culture growth conditions *after* the irradiation (Figure 7-1). This phenomenon, known as liquid holding recovery, will be treated in more detail in Section 7-6,B.

The type of recovery as well as the extent of recovery will depend quite obviously on the nature of the molecule that has been damaged. Thus, in studies on division inhibition and immobilization of paramecia by UV irradiation it was found that:

(a) The action spectrum for division delay implicated nucleic acid whereas that for immobilization paralleled the absorption spectrum for an albumin-like protein (see Figure 1-12).

(b) Most of the organisms irradiated at 226 nm recovered mobility whereas none of those that were immobilized by 267 nm irradiation recovered, unless photoreactivating conditions (Section 7-3,C) followed the irradiation. The evident explanation is that damage to proteins may be circumvented if the genetic information for the replacement of these proteins is intact, but that damage to the functional regions in the primary genome may require direct repair of the damage if recovery is to be observed.

Eventually we may understand the molecular bases of recovery mechanisms in the cells of higher organisms. However, most of our current understanding comes from studies with the simplest systems, the bacteria and their viruses. Much of the ensuing discussion will therefore be concerned with these relatively simple systems in

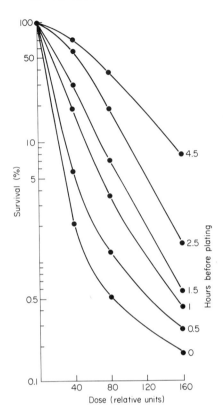

FIGURE 7-1.   Ultraviolet survival curves for *E. coli* B. After irradiation the cells were suspended in a liquid medium without an energy source for the times indicated before being plated on nutrient agar. It is evident that both the slopes and the shapes of the survival curves can be altered by the postirradiation treatment of the cells. [From R. B. Roberts and E. Aldous, *J. Bacteriol.* **57**, 363 (1949).]

the hope that generalizations to more complex and highly evolved living systems may ultimately be made.

Before discussing specific repair mechanisms it may be helpful to review the major photoproducts known to have biological effects. These photoproducts have been determined by the examination of DNA that has been extracted from irradiated cells. (Figure 7-2.) Some of these effects (e.g., denaturation) can be seen as secondary effects resulting from photoproducts such as dimers and hydrates.

Others, like strand breaks, can occur indirectly as a consequence of the dark repair mechanism to be discussed in Section 7-4.

## 7-2. FORTUITOUS RECOVERY

### A. Biologically Undetectable Damage

By way of clarifying the definition of repair as related to recovery in general, we should dispense with some of the modes of cellular recovery which have very little to do with the direct repair of photochemical damage. It is sometimes possible for an organism to survive damage in its genetic material without any "conscious" recognition that the damage is present. The most obvious example would be the case in which the damage is irrelevant to the effect being measured. Thus, as discussed in Section 4-4, UV-damaged cytosine may behave in an RNA polymerase system *in vitro* as though it were uracil. However, depending on its position in the codon triplet, such a transition might or might not result in an amino acid change in some resultant protein. Furthermore, damage to regions

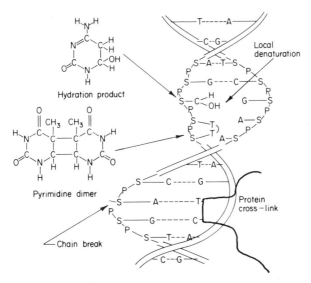

FIGURE 7-2. Schematic illustration of the various alterations found in DNA extracted from cells that have been irradiated with ultraviolet light. [Adapted from R. A. Deering, *Sci. Am.* **207**, 135 (1962).]

of the genome that are not being actively transcribed may be of little consequence to the organism until such time as the information in those regions is required.

### B.  Polyploidy

In this general category might be included any inactivation of units of function for which a redundancy exists in the cell. Thus, for example, the inactivation of a particular enzyme molecule would have no detectable effect on cell growth if there were many such enzyme molecules present in the cell. The effect of polyploidy on X-ray survival is quite effectively illustrated in a comparison of the survival curves for haploid and diploid strains of yeast: the haploid form exhibits the simple exponential survival curve expected for a single target inactivation process while the diploid strain gives rise to a multicomponent inactivation curve with an extrapolation number of roughly 2, as consistent with two sensitive targets per cell. Another example is seen in the case of the UV inactivation of chloroplast duplication in the unicellular green alga, *Euglena gracilis*. All 30 proplastids must be "hit" to permanently inactivate the ability of this organism to produce chloroplasts (Section 6-4). It is important to realize that inactivation curves with shoulders may indicate either polyploidy or the presence of repair mechanisms, and that it may be quite difficult to ascertain which effect is responsible. The correlation of an apparent polyploidy with cytological observation sometimes can resolve this question, as in the case of *Euglena* chloroplast inactivation.

### C.  Multiplicity Reactivation

This phenomenon, first observed by Luria, involves the cooperative effects of UV-inactivated bacteriophage to produce some viable phage when the host cell is multiply infected (Section 6-7 and Figure 6-13). Multiplicity reactivation (MR) has also been demonstrated with animal viruses, and it has been speculated that it may even occur between nuclei within uninfected diploid cells. MR has also been observed in phage after deleterious treatments other than UV, such as X-rays, gamma rays, nitrous-acid, and $^{32}$P decay. The phenomenon evidently involves genetic recombination in which the random process of molecular rearrangement may result in the production of a viable genome from the undamaged components of otherwise nonviable genomes. Although the recombinational event is not directly related to the presence of UV photoproducts, there is an indirect effect in that the damage which leads to excision-repair also stimulates genetic recombination. Thus, the nonspecific stimulation of genetic recombination may increase the probability of MR in UV-damaged phage genomes. This stimulation of recombination by damage to DNA has also been shown for low energy X-rays on T4 phage. The cooperation between damaged genomes need not involve physical exchange of DNA. For example, cooperation at the level of gene function has been demonstrated. UV-inactivated phage were shown to be able to contribute some critical function essential to the reproduction of other UV-damaged phage which lacked this function. The relative contributions of genetic exchange and functional cooperation to the phenomenon of MR remain unknown.

## D.  Cross Reactivation (Marker Rescue)

The process known as cross reactivation or marker rescue is essentially the same as the molecular rearrangement aspect of multiplicity reactivation. The bacteria are infected with two genetic types of phage, of which one has been UV irradiated. Genetic markers from the UV-inactivated phage may be physically incorporated into the genome of the unirradiated phage. This rescue of genetic markers can be demonstrated even after most of the genetic information in the "donor" phage particles has been destroyed by irradiation.

## E.  Suppression of Prophage Induction

It has been suggested that the UV sensitivity of bacteria may be enhanced by the presence of a UV-inducible prophage. Any condition that might inhibit the induction of such a prophage would then fortuitously lead to increased resistance of the bacteria to irradiation (Section 6-8). It is clear that such an apparent recovery factor might bear no relation to the repair of potentially damaging photoproducts in either the prophage or the bacterial genome.

7-3. REVERSAL OF DAMAGE BY REPAIR *in Situ*

*A. Decay of Photoproducts*

Perhaps the simplest mechanism of repair is the one that involves the spontaneous reversion of photoproducts to the original undamaged state. Obviously the cell can have little control over this sort of restoration, but environmental conditions can have a great deal to do with it. The hydration products of the pyrimidines are known to revert spontaneously. Also, several of the dimeric thymine photoproducts have been shown to be reversed by acid catalysis.

The half-life for the UV-induced hydrate of cytidylic acid is 48 minutes at 25°C and pH 5.7. Another cytosine photoproduct, as yet uncharacterized, has been shown to have a half-life of only 8 minutes at 0°C and pH 6.5. This emphasizes an important and troublesome fact in photochemical studies with biological systems. Perhaps many such photoproducts are formed within irradiated cells but they may have such a fleeting existence that the biological effect must occur instantaneously at the point of function before the photoproduct decays. This explanation would be consistent with the general observation that cells are more sensitive to UV if they are actively growing (e.g., replicating DNA) than if they are not replicating DNA.

Thermal reactivation, as originally reported by E. H. Anderson, may in part involve the increased decay rate for labile photoproducts at higher temperatures. The survival of UV-irradiated *E. coli* B is higher if the cells are incubated for several hours at 45°C immediately after the irradiation and allowed to grow at 37°C than if they are only incubated at 37°C. It has been found that a delay in the exposure of the irradiated cells to the elevated temperature results in a decreased amount of thermal reactivation and after an elapsed time of three hours no thermal effect can be seen. Thermal reactivation is observed only following irradiation with wavelengths shorter than about 313 nm, and the effect on *E. coli* B is much more pronounced than that on strain B/r. In fact, incubation of irradiated *E. coli* strain B at 45°C results in a survival curve nearly identical to that obtained for strain B/r. The fact that a small thermal reactivation effect is also observed in *E. coli* strain $B_{s-1}$, which is deficient in the excision-repair mechanism (Section 7-4), indicates that the effect is not principally one of enhancement of the excision-repair mode at higher temperatures. This reactivation mechanism may in part relate to the division-stimulating effect of elevated temperatures which will be considered in more detail in Section 7-6,G.

Another reactivation phenomenon which may involve an enhanced decay of photoproducts is that of catalase reactivation. A mutant of *E. coli* which was unable to synthesize catalase because of a hemin deficiency was found to have increased UV resistance when hemin was added to the growth medium. The addition of hemin to the wild type culture had no effect on UV survival. It is thought that the peroxidase activity of the catalase is involved in the reactivation effects. It is not at all clear whether photoproducts in DNA are reduced or whether the effect involves the reduction of UV-induced peroxides that might act indirectly in the cell.

We can say little more about the relevance of photoproduct decay to biological recovery until we understand more about the kinds of photoproducts that can revert spontaneously. Nevertheless, we should be aware of this possible mode of recovery, particularly when we consider environmental effects on cellular survival.

## B.   Direct Photoreversal of Pyrimidine Dimers

The maximum yield of thymine dimers in UV-irradiated DNA is dependent upon the wavelength of the irradiation as well as upon the conditions of irradiation. After large enough UV doses at a given wavelength a photosteady state is attained in which the relative number of dimers and free thymines does not change (Section 4-5). Thus, dimer formation is a reversible process and the dimers may be reverted to free thymines by UV absorption. At 275 nm the steady state yields 20% of the thymine as dimers in phage T4 DNA while at 235 nm this equilibrium yields only 1.7% of the thymine as dimers. The absorption for dimers is roughly 1000 times greater at 239 nm than at 280 nm (Figure 4-7). After UV inactivation of transforming principle DNA by irradiation at 280 nm, partial recovery by reirradiation at 239 nm has been observed. It has been estimated from such studies that more than 50% of the loss of biological activity can be ascribed to thymine dimers in the transforming DNA. It was found, however, that prolonged irradiation at the shorter wavelength eventually reduced the survival level of genetic markers in the DNA from that obtained after maximum reversal of the 280 nm effects. This indicated that the 239 nm radiation was producing some additional types of damage.

This direct photoreversal process has been very useful in studies on transforming DNA but it is not of significance as a repair mechanism in living systems. The problem is that short wavelength irradiation itself promotes an equilibrium level of pyrimidine dimers which can not be tolerated in cellular DNA. Even though currently existing organisms appear unable to use this mode of recovery, it might well have been important in primitive photochemical recovery phenomena and it may have been the chemical reaction which finally evolved into the enzyme-catalyzed form of photoreactivation.

## C. Enzyme-Catalyzed Photoreactivation

*C1. Mode of Action.* The most thoroughly characterized cellular recovery mechanism is that of enzymatic photoreactivation, in which illumination with visible light facilitates the direct repair *in situ* of photoproducts produced by UV in DNA. It is at first puzzling to consider that the phenomenon actually involves the use of 3 ev photons to undo the damage promoted by 5 ev photons. However, much of the mystery of the process began to be dispelled with the isolation of cellular extracts that could promote the process *in vitro*. The principal effect of enzymatic photoreactivation has been clearly shown to involve the splitting of pyrimidine dimers *in situ*. Transforming DNA exposed to 280 nm radiation and subjected to maximal enzymatic photoreactivation can not be further reactivated by subsequent exposure to 239 nm radiation. Transforming DNA subjected only to 239 nm radiation may be enzymatically photoreactivated, since 239 nm both produces and splits pyrimidine dimers (Section 7-3, B). These results strongly implicate the reversal of dimers as the principal mechanism of enzymatic photoreactivation. The direct chemical conversion of thymine dimers to free thymine by the photoreactivating enzyme has also been shown.

*C2. Properties of the Enzyme.* Evidence for the process of photoreactivation was first reported by Whitaker in studies with *Fucus* eggs. However, the real beginnings of the study of photoreactivation stem from the nearly simultaneous rediscovery of the effect in bacteria and bacteriophage by Kelner and Dulbecco, respectively.

Cellular extracts from a number of biological systems have been shown to possess photoreactivating activity. Most extensively characterized have been those from *E. coli* and Baker's yeast. Extracts from Baker's yeast have been preferred to those from *E. coli* because the latter requires a dialyzable component in the reaction mixture. The chemical specificities of the two systems would seem to be identical, however, as judged by the fact that DNA maximally repaired by one system is not further repaired by the other system.

Photoreactivation with the yeast extract system follows classical enzyme kinetics. The enzyme binds specifically to UV-irradiated (but not unirradiated) DNA to form a complex that is stable in the dark. The existence of this complex in the dark is supported by the fact that the enzyme becomes stabilized to heat inactivation (a common property of enzymes with bound substrates) and follows the DNA in sedimentation and gel filtration. If the complex is illuminated with visible light it separates into the active enzyme component and a repaired DNA that can no longer bind to the enzyme. Illuminating the enzyme and/or damaged DNA prior to combining the two has no effect on either binding or repair of UV damage. Reciprocity has been demonstrated for the light reaction both *in vitro* and *in vivo*. The enzyme has now been highly purified to a 6000-fold enrichment by Muhammed but as yet no component has been identified with an absorption spectrum that resembles the action spectrum for *in vitro* photoreactivation and the chromophore remains unknown. One might suppose that the relevant chromophore occurs as a result of the complexing of enzyme and DNA. This possibility should be examined.

The specificity of the photoreactivating enzyme has been extensively studied by competition of various UV-irradiated substrates with the enzymatic repair of a UV-irradiated transforming principle DNA. Thus, it has been shown that only polynucleotide strands containing adjacent pyrimidines are photoreactivable. As expected, if pyrimidine dimers are photoreactivable lesions, the alternating copolymer d(AT):d(AT) does not compete but the homocopolymer dA:dT does. The fact that poly dG:dC does compete was the first indication that damage other than thymine dimers could be repaired by this enzyme. It has since been shown that all combinations of the pyrimidines can form dimers and that these can be eliminated from DNA by photoreactivation, although thymine dimers are eliminated more efficiently than are the other types. No types of DNA damage other than those produced by UV have been shown to be photoreactivable. Competition studies indicate that the photo-reactivating enzyme can bind to 5-bromouracil-containing DNA but that it is not subsequently released after illumination of the complex. The competing irradiated DNA need not be double stranded; UV-irradiated DNA from the single-stranded DNA bacteriophage ΦX174 will compete and even unirradiated single-stranded DNA will compete to a small extent that is not further diminished by pretreatment with the photoreactivating enzyme. No competition is provided by RNA or ribonucleotide homopolymers whether irradiated or not. The minimum size of the polydeoxyribonucleotide substrate for the enzyme would appear to be about nine nucleotides in order for binding to the enzyme to occur. A chain of oligodeoxy-thymidylates 18 residues long competed as well as a chain several hundred residues long, while a chain containing nine residues competed only slightly and one of eight residues not at all. An action spectrum for photoreactivation of *Hemophilus influenzae* transforming DNA by the *E. coli* extract was obtained by Setlow and Boling, and it indicated peaks at 360 nm and 390 nm, (Figure 7-3), corresponding roughly to the broad peaks obtained in an action spectrum for photoreactivation of *E. coli* B/r *in vivo* (Figure 7-4).

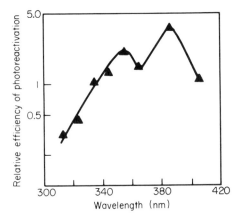

FIGURE 7-3.   Action spectrum for photoreactivation of UV-irradiated transforming DNA from *Hemophilus influenzae* with yeast extract. [From J. K. Setlow and M. E. Boling, *Photochem. Photobiol.* **2**, 471 (1963).]

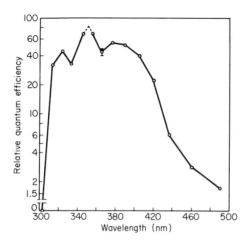

FIGURE 7-4.   Action spectrum for photoreactivation of *E. coli* B/r viability following UV irradiation. [From J. Jagger and R. Latarjet, *Ann. Inst. Pasteur* 91, 585 (1956).]

It has been suggested that a third peak (324 nm) in the *in vivo* action spectrum not present in the *in vitro* spectra may involve an indirect photoreactivation effect (Section 7-6,D).

A nonphotoreactivable mutant of *E. coli* B has been isolated by Harm and Hillebrandt. Its extract lacks the ability to photoreactivate transforming DNA and this strain does not lose pyrimidine dimers from its DNA upon exposure to visible light.

It is of interest, however, that a number of UV-induced mutations in this nonphotoreactivable strain were found to be photoreversible. This led to the suggestion that there may be two enzymes involved in photoreactivation, one which breaks pyrimidine dimers and the other which deals with some mutagenic photoproduct that is not a pyrimidine dimer. The 324 nm peak in the action spectrum for *in vivo* photoreactivation in *E. coli* B/r could conceivably involve such a second enzyme.

*C3.   Generality of the Process in Vivo.* Enzymatic photoreactivation has been demonstrated in a wide variety of cellular systems. It has also been shown to reverse a large number of different biological effects. Thus, the demonstration of photoreversibility of UV mutagenesis soon followed the original discovery of the phenomenon in relation to UV killing of bacteria. As now understood from our knowledge of the molecular mechanism of the process and the particular photoproducts that are accessible to repair, it is not surprising to find that the photoreactivable biological damage is almost invariably related to some DNA function. Thus, the photoreactivation of UV-induced division delays in *Paramecia* are now explainable in terms of DNA repair. Of particular interest in retrospect are the photoreactivation effects in cytoplasmic organelles. The photoreactivation of "petite" mutants in yeast and UV-bleached mutants of *Euglena gracilis* was shown even before direct evidence was obtained for the presence of DNA in mitochrondria and chloroplasts, respectively. The UV induction of a temperate bacteriophage has been shown to be photoreversible, which indicates that DNA is the principal target for the UV induction process.

There is a growing list of organisms that have been shown to contain the photoreactivating enzyme. However, equally significant is the list of those cell types that do not contain this enzyme. The very simplest free-living cells, the pleuropneumonia-like organisms (PPLO) do photoreactivate UV damage, thus emphasizing the importance of the process to survival, since the genomes for these cells contain only enough information for about 700 different proteins. To list a number of other systems that show a positive response to photoreactivating light: sea urchin eggs, marsupials such as the woolly possum or the Australian kangaroo, fibroblasts and embryonic chicken tissues, chicken brain, African clawed toad, cell lines from fish such as the Grunt fin, protozoans such as *Tetrahymena*, etc. However, photoreactivation has not yet been demonstrated in mammalian tissues (e.g., HeLa cells or hamster kidney cells). Finally, it is of interest that the radiation-resistant bacterium, *Micrococcus radiodurans*, does not seem to possess a photoreactivation enzyme. This raises an important point with regard to detection of photoreactivation *in vivo*. In systems with very efficient dark repair mechanisms it is conceivable that photoreactivation would not be detectable.

Photoreactivation of RNA-containing bacteriophage and the RNA of plant viruses such as TMV has also been reported, although it is possible that this is not due to the sort of enzymatic mechanism discussed above but rather to indirect effects as discussed in Section 7-5.

## 7-4. RECONSTRUCTION OF DAMAGED DNA

### A. Evidence for Excision-Repair

The studies of R. B. Setlow and co-workers provided the first experimental evidence leading to a model for repair of UV-damaged DNA in the dark. It was shown that UV had a much greater effect on DNA synthesis in the sensitive strain *E. coli* $B_{s-1}$ than in the resistant strain *E. coli* B/r. Since it was found that the numbers of thymine dimers imposed on the two strains by a given UV dose was the same, it seemed evident that the resistant strain could somehow remove or bypass these apparent blocks to replication. The mechanism for this recovery was clarified when it was shown that the resistant strain (but not the sensitive strain) released thymine dimers from its DNA during subsequent incubation in the dark after irradiation. Similar results were soon reported by Boyce and Howard-Flanders for resistant and sensitive strains of *E. coli* K-12. A repair mechanism was postulated in which defective regions in one of the two DNA strands could be excised and then subsequently replaced with normal nucleotides, utilizing the complementary base pairing information in the intact strand. This mechanism (Figure 7-5), which has come to be known colloquially as "cut and patch", has turned out to be of widespread significance for the repair of a variety of structural defects in DNA. The existence of this mechanism also provides a logical explanation for the evolution in most organisms of two-stranded DNA, which comprises a redundancy of information.

Direct physical evidence for the repair replication or "patch" step in the postulated scheme was provided by the studies of Pettijohn and Hanawalt. These studies began with attempts to isolate partially replicated fragments of the bacterial chromosome by imposing blocks to replication (i.e., UV-induced pyrimidine dimers). Replication was followed by the use of the thymine analog, 5-bromouracil (5BU), as a density label in newly synthesized DNA, and by the subsequent analysis of the density distribution of isolated DNA fragments in a cesium chloride density gradient (Figure 7-6). This is essentially the method developed by Meselson and Stahl and utilized by them to prove that DNA normally replicates semiconservatively. When 5BU was used to label the DNA synthesized after UV irradiation of *E. coli* strain TAU-bar to $10^{-2}\%$ sur-

I. Recognition

II. Incision

III. Excision

IV. Degradation

V. Repair replication

VI. Rejoining

Alternate Steps

III.' Repair replication

IV.' Excision

V.' Degradation

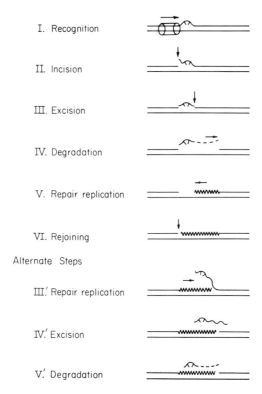

FIGURE 7-5. Schematic representation of the postulated steps in the excision repair of damaged DNA. Steps I through VI illustrate the "cut and patch" sequence. An initial incision in the damaged strand is followed by local degradation before the resynthesis of the region has begun. In the alternative "patch and cut" model the resynthesis step III' begins immediately after the incision step II and the excision of the damaged region occurs when repair replication is complete. In either model the final step (VI) involves a rejoining of the repaired section to the contiguous DNA of the original parental strand.

FIGURE 7-6. Protocol for the demonstration of normal replication and repair replication of DNA in growing cells.

The DNA is first radioactively labeled (e.g., by the incorporation of carbon-14 labeled thymine); then the cells are permitted to incorporate another radioactive label at the same time that a "density label" is being incorporated (e.g., hydrogen-3 labeled 5-bromouracil). 5-Bromouracil (5BU) is an analog of thymine that can be incorporated into DNA in place of the natural base thymine. Since 5BU is more

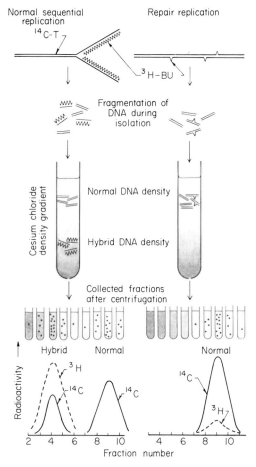

dense than thymine it has the effect of increasing the density of the DNA fragments that contain it. This density increase is, of course, proportional to the relative amount of thymine and 5BU in the DNA. The density distribution of the isolated DNA fragments is analyzed by means of equilibrium sedimentation in the ultra-centrifuge in a density gradient of cesium chloride solution. At equilibrium the DNA fragments will be found in the gradient at positions that correspond to their buoyant densities rather than to their size. This is essentially the method developed by Meselson and Stahl and utilized by them to prove that DNA normally replicates semiconservatively (shown on left half of figure). Parental DNA fragments that contain short regions of repair may differ little in density from those that contain no 5BU (shown on right half of figure). [Adapted from P. C. Hanawalt and R. H. Haynes, *Sci. Am.* **216**, 36 (1967).]

vival, the density pattern observed was not as expected for normal semiconservative replication. Instead of a hybrid density band in the gradient, the initial incorporation of the 5BU label after UV resulted in no detectable shift in density from the normal parental DNA band.

Proof that the *early* incorporation of 5BU into DNA fragments is the postulated step of repair replication has come from a number of control experiments as follows:

(a) This mode of replication is not observed if the bacteria are illuminated with visible light to allow the *in situ* photoreactivation of pyrimidine dimers prior to 5BU labeling.

(b) It is not observed following UV irradiation of the UV sensitive strain *E. coli* $B_{s-1}$, which is unable to perform the excision step in the repair sequence.

(c) The nonconservative mode of repair replication can also be demonstrated by the use of $D_2O$, $^{13}C$, and $^{15}N$ as density labels for newly synthesized DNA to rule out possible artifacts caused by the pathogenicity of 5BU.

(d) In low dose experiments (in which viability was as high as 80%) it was demonstrated that DNA which had incorporated 5BU nonconservatively after UV irradiation could then proceed to replicate by the normal semiconservative mode.

## B.   The Steps in Excision-Repair

*B1.   Recognition.* The first step in the repair process must involve the recognition of the damaged region in the DNA. The photoreactivating enzyme, of course, has been shown to be capable of recognizing pyrimidine dimers. However, unlike the photoreactivation system, the excision-repair system is capable of recognizing a variety of structural defects in DNA which do not involve pyrimidines and which do not result from UV effects. Repair replication has now been observed following treatment of bacteria with the bifunctional alkylating agent, nitrogen mustard, which primarily attacks the 7-nitrogen position of guanine. It has also been demonstrated following exposure of bacteria to the powerful mutagen, nitrosoguanidine. Indirect evidence that still other DNA damage can be recognized and repaired comes from the finding that mitomycin C treatment leads to DNA degradation in UV-resistant bacteria but not in UV-sensitive strains. Thus, it may not be the precise nature of the base damage that is recognized but rather some associated secondary structural alteration in the phosphodiester backbone of the DNA. The damage recognition step may be formally equivalent to threading the DNA through a close-fitting "sleeve" that gauges the closeness-of-fit to the Watson-Crick structure.

The repair of damage which leads to single strand breaks in the DNA, such as produced by decay of incorporated $^{32}$P, by X-irradiation or by treatment with methylmethanesulfonate, might not require the incision step in the process and could proceed to the excision step (Figure 7-5). This may explain why some mutants sensitive to UV (presumably because of a deficiency in the recognition step and/or incision) are resistant to methylmethanesulfonate.

The cell must usually recognize and repair damaged DNA before the damaged section is replicated. One would suspect that there might be some difficulty in repairing a pyrimidine dimer once the DNA strands had begun to separate to facilitate semiconservative replication. Some mechanism would seem to be necessary to halt normal replication until repairs had been made. Two possibilities might be suggested, although there is at present no experimental evidence for either one.

(1) The clamping of the recognition enzyme to the damaged region of DNA may provide a physical constraint to the rotation of the molecule necessary for the unwinding of strands prior to normal replication.

(2) Distortions in the secondary structure of the DNA may alter the pattern of "breathing" of the DNA in which regions of partial denaturation are believed to travel along the structure. A pyrimidine dimer might alter the breathing pattern and result in a "message" being transmitted to the site of normal replication to halt synthesis.

*B2. Incision.* Following the recognition of damage in DNA, a necessary prerequisite to the excision of the damaged region is the incision step or production of a single strand break near the damage. The incision step may precede the excision step although it has not been ruled out that the two might normally occur as a single enzymatic process.

The incision step has been demonstrated in cell-free extracts of *Micrococcus lysodeikticus* by Rörsch and co-workers in an elegant series of experiments with the double-stranded form of bacteriophage ΦX174. This so-called replicative form, when irradiated, can be repaired in spheroplasts of wild type *E. coli* but not in mutants defective in the recognition (and incision?) step in excision-repair. However, a marked increase in biological activity was observed when the damaged DNA was first incubated in an extract of *Micrococcus lysodeikticus* before infection of the spheroplasts. Confirmation that the extract was indeed performing the incision step in repair came from studies on the sedimentation behavior of untreated and UV-irradiated phage DNA after exposure to the extract. The actual excision of pyrimidine dimers from UV-irradiated DNA in *M. lysodeikticus* extracts has also been shown.

*B3.   Excision and Repair Replication.* The processes of excision and replacement of damaged nucleotides may occur as separate steps or they may be carried out concurrently with a peeling back of the defective DNA strand. It is thought that a limited DNA breakdown occurs in the course of removal of the damage, but it is very difficult to obtain a reliable estimate of the size of the excised segment in relation to the number of defects repaired. First, it is difficult to know how many defects have been repaired. And second, it is possible that some more severely damaged cells may be undergoing nonspecific DNA degradation that would increase the estimate. Most of the DNA breakdown is to the mononucleotide level. However, the released pyrimidine dimers in irradiated bacteria are found as part of tri- and tetranucleotides. It is quite conceivable that the dimers were excised in longer segments but that these were subsequently reduced by nuclease degradation. It should be pointed out that no direct correlation has been made between the observed DNA degradation and the specific enlargement of regions from which dimers have been excised. The method of Pettijohn and Hanawalt might yield an estimate for the size of the repaired region but there are problems here also. For example, the 5BU incorporated during repair replication must compete with released thymidine from the damaged DNA.

The known specificities of Exonuclease III and the DNA polymerase from *E. coli* make these enzymes attractive candidates for excision and repolymerization, respectively. An *in vitro* model for the "cut and patch" process was demonstrated in which a portion of one strand of a transforming DNA was degraded with Exonuclease III with the concomitant loss in biological activity. The biological activity was subsequently restored by the action of the DNA polymerase.

The fact that thymine deprivation does not interfere with dimer excision would tend to favor the "cut and patch" model for excision-repair. On the other hand, the release of dimers *in vivo* appears to be an energy dependent process: in the absence of glucose the rate of DNA solubilization in UV-irradiated bacteria is sharply reduced. None of the known nucleases that attack DNA require an energy source for their activity. This, then, might be considered evidence for the close coupling of the excision and the replacement steps, since the repair replication must require an energy source as does normal replication. Kornberg and co-workers have demonstrated exonucleolytic activity in highly purified preparations of *E. coli* DNA polymerase. A single-strand break in a double-stranded DNA template is translated along the structure as nucleotides are released from the 5'-phosphate end while the polymerase adds nucleotides to the 3'-hydroxyl end. These results support the "patch and cut" model of repair (Figure 7-5): the simultaneous excision and replacement of nucleotides.

Two lines of evidence support the concept that different enzyme systems are involved in the normal semiconservative mode and in the repair mode of DNA synthesis in *E. coli*. Firstly the repair mode of synthesis is essentially unaffected at the restrictive temperature in several DNA synthesis deficient temperature-sensitive mutants. (Such mutants synthesize DNA and grow normally at 35°C but normal replication stops when the temperature is raised to 42°C, presumably because some component in the replicase complex is thermosensitive.) Secondly, it has been shown that the repair mode exhibits a greater selectivity for thymine over 5BU than the normal mode, when both the natural base and its analog are present in the culture medium. Since it can be presumed that both modes utilize the same internal pool of nucleoside triphosphate precursors the repair polymerase seems to have a more stringent requirement for thymine than does the normal replicase.

*B4. Rejoining.* The excision-repair process is completed by the rejoining of the repaired segment to the continuous intact DNA strand to restore the integrity of the two-stranded molecule. The best evidence for the occurrence of this step *in vivo* was found by the examination of molecular weights of the single-stranded DNA by sedimentation in alkaline sucrose gradients following gentle lysis of bacteria on top of the gradients (method of McGrath and Williams, Figure 10-7). Thus, large single-stranded DNA fragments were obtained from normal cells and smaller pieces were seen shortly after irradiation. A subsequent reduction in the number of strand breaks with time during incubation after irradiation could be followed by this method. R. B. Setlow has shown that the breaks occur only in the *damaged* strand. The recently isolated poly-nucleotide ligase which performs a single-strand rejoining function *in vitro* may well be the enzyme responsible for the rejoining step in repair.

## C. *Generality of Excision-Repair*

Evidence for the excision-repair scheme has been obtained in quite a number of microorganisms in addition to *E. coli*. Of par-ticular interest is the finding of an extremely efficient excision mechanism in the highly radioresistant bacterium *Micrococcus radiodurans*. Excision-repair has also been demonstrated in the smallest living cells, the mycoplasma, by D. Smith. The presence of a DNA repair mechanism in these cells attests to the general importance of such mechanisms for the maintenance of viability in even the simplest organisms.

Most attempts to demonstrate excision of UV-induced pyrimi-dine dimers from DNA in mammalian cells have been unsuccess-ful. Thus, Klimek reported the formation of thymine dimers in the DNA of mouse L-cells in culture but he could not demonstrate their release from DNA during postirradiation incubation. The problem may be one of the recovery of the released segments: they might be so large that they are acid insoluble. Regan and Trosko were able to demonstrate the preferential removal of thy-mine dimers from three human cell lines (RA, RAX10, and HeLa) in tissue culture. Correspondingly, Rasmussen and Painter found an "unscheduled" DNA synthesis stimulated by UV in cultures of HeLa cells. Their analysis of the replicated DNA by the 5BU den-sity labeling method has provided support for the interpretation that repair replication is occurring in this system.

In studies on repair replication in human skin fibroblasts Cleaver made an important discovery that correlates carcinogenesis with defective repair of DNA. Fibroblasts from patients with Xeroderma pigmentosum were found to exhibit much reduced levels of repair replication. It was suggested that the failure of DNA repair might be related to the fatal skin cancers that patients with this hereditary disease develop upon exposure to sunlight.

An improvement in the method of Pettijohn and Hanawalt has been utilized by Brunk to prove that the protozoan, *Tetrahymena pyriformis* can perform repair replication following UV irradiation. The method has the advantage that it is based on the examination of DNA that has also replicated normally. The 5BU density label is used but replication is allowed to proceed until significant amounts of hybrid, normally replicated DNA are evident in CsCl gradient analysis. The hybrid DNA is collected and denatured (to separate the light parental strand from the heavy daughter strand) in a second *alkaline* CsCl gradient and the amount of 5BU label in the parental strand is taken as a measure of repair.

The application of the bacterial excision-repair system to the DNA of infecting UV-irradiated bacteriophage was termed host-cell reactivation (HCR) long before the mechanism was understood. It was discovered by Garen and Zinder who found that UV or X-irradiation of *Salmonella* host cells prior to infection with UV-irradiated phage P22 led to lower survival for the phage. An hypothesis was proposed to explain the phenomenon assuming that homologous regions of the bacterial genome could replace damaged genetic material of the phage. Serious doubt as to the validity of this explanation was raised by the finding that the phages T1, T3, and T7 were much more sensitive to UV when grown on *E. coli* strain $B_{s-1}$ than on the more resistant strain B. This discovery was followed by the isolation of other bacterial mutants which did not reactivate infecting phage and by the localization of the mutated sites on the genetic map. The sites involved with HCR were often shown to affect the UV sensitivity of the host cells themselves. These genetic studies preceded the molecular investigations which elucidated the steps in the excision-repair process. It has been found, however, that some UV-sensitive bacterial strains are still able to perform HCR. Thus, strain $B_{s-2}$ is almost as sensitive to UV as strain $B_{s-1}$ but it still exhibits HCR. The possibility of reduced repair efficiency in this strain has been suggested (i.e., the residual repair may be sufficient for the phage but not for the host). As discussed in Section 7-4,B, the *in vitro* study of the excision-repair process has been carried out largely with HCR systems. The expectation that only double-stranded DNA phage can be reactivated by this mechanism was supported by studies in which the difference in UV sensitivity of single-stranded ΦX174 DNA and the double-stranded replicative form were compared. Further confirmation was obtained by Sauerbier, who took advantage of the selective inhibition of HCR by caffeine to demonstrate that HCR of intracellularly irradiated ΦX174 can only occur from about three minutes after the onset of phage development, by which time the replicative form has been synthesized.

The incorporation of 5BU into the phage DNA has been shown to inhibit HCR although no effect is seen if the 5BU is in the host DNA. This was early evidence against the recombination model for HCR, as was the finding that phage damaged by nitrous acid could be reactivated by multiplicity reactivation but not by HCR. The mechanism by which 5BU interferes with excision-repair is not understood. However, extensive breakdown of 5BU DNA following UV irradiation of bacteria has been reported. The reported reduction of HCR by the UV irradiation of the host cell is most logically explained by the competition of damaged host DNA with damaged phage DNA for a limited number of repair complexes in the host cell.

### 7-5.  DNA Synthesis on Unrepaired Templates

Mutant strains of *E. coli* have been isolated which exhibit reduced genetic recombination in bacterial mating experiments and are also unusually sensitive to UV. This led to the suggestion that genetic recombination might have some steps in common with the excision-repair of damaged DNA. However, recombination deficient (REC⁻) mutants were found to be fully capable of excision and repair replication. Howard-Flanders and co-workers showed that double mutants, deficient in *both* recombination and excision, were more sensitive to UV than either type of single mutant (Figure 7-7). This suggested that there might be a dark-repair mechanism that operated in addition to the excision-repair mode. The details of such a mechanism are not yet clear but the study of DNA synthesis in excision-deficient bacterial mutants is supplying some clues. Rupp and Howard-Flanders found that the newly-synthesized DNA in UV-irradiated excision-deficient bac-

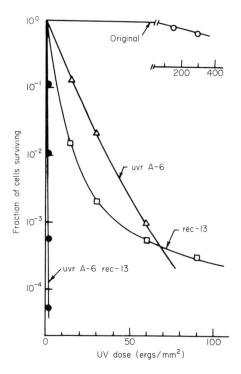

FIGURE 7-7. The sensitivity of colony forming ability to ultraviolet light in several UV-sensitive bacterial mutants. The mutant strain uvr-A6 is unable to excise thymine dimers. The mutant rec-13 is defective in genetic recombination. The double-mutant uvr-A6 rec-13 is deficient in both excision and recombination and it is more sensitive than either single mutant. [Adapted from P. Howard-Flanders and R. P. Boyce, *Radiation Res.* Suppl. 6, 156 (1966).]

teria existed in mean lengths that corresponded to the expected distances between dimers. Zone sedimentation in alkaline sucrose gradients (method of McGrath and Williams, see Figure 10-7) was used to show this. Thus, it appeared that pyrimidine dimers did not stop synthesis but that they caused gaps to be left in daughter strands as synthesis slowly proceeded past them. Examination of the DNA at later times indicated that the gaps had been repaired. K. C. Smith and co-workers have shown that in REC⁻ strains the gaps are not repaired. The role of the REC function in this repair scheme is thus supported.

7-6.  PHYSIOLOGICAL AND ENVIRONMENTAL EFFECTS ON
      REPAIR AND RECOVERY

A.  New Repair Mechanisms (or Indirect Effects on
    Known Ones)

Of the various treatments that alter sensitivity of cells to killing
by photons, some have been shown to affect specific repair mechanisms such as enzymatic photoreactivation or excision-repair. Other
effects on cell survival may involve additional repair mechanisms
yet to be elucidated. It is likely, however, that many of these treatments simply enhance repair processes or suppress factors that
inhibit recovery. It might be predicted for example that excision-
repair would be enhanced by factors that inhibit normal DNA replication. This principle is consistent with most of the observations
discussed in the following sections.

B.  Liquid Holding Recovery

If UV-irradiated cells of E. coli B are allowed to stand in buffer
for a period following irradiation before being plated on nutrient
agar they show an increased viability over cells which were plated
immediately after irradiation (see Figure 7-1). This phenomenon,
known as liquid holding recovery (LHR), would seem to be explainable in terms of the effect of a growth lag on the efficiency of the
excision-repair mechanism. Maximum enhancement of survival
is obtained after about 4–6 hours in liquid at room temperature. The
overlap of LHR with photoprotection (see Section 7-6,E) indicates
that the two mechanisms modify the same lesions. Cells allowed
optimal recovery by one of these treatments showed no further
recovery when subjected to the other. Furthermore, overlap of
LHR with photoreactivation in E. coli B has been demonstrated.
It was concluded that LHR acts upon photoreactivable damage,
thus implicating pyrimidine dimers and indirectly implicating the
excision-repair scheme. A more direct correlation with excision-
repair was the finding that the strain $B_{s-1}$ showed almost no LHR.

A number of UV-sensitive mutants of E. coli K-12 which exhibit
LHR have been identified as recombination deficient. Further, it
has been demonstrated that LHR requires the same gene function
that controls excision but that the observation of LHR depends
upon the presence of certain REC⁻ mutations. Since the rec and uvr

genes appear to control two distinct types of repair (Section 7-5), it is possible that LHR can only be observed when the $REC^+$ mode of repair is absent.

## C.  UV Sensitivity and the DNA Replication Cycle

The striking changes in UV sensitivity that accompany unbalanced growth may also be due to growth inhibition effects, more specifically to the inhibition of DNA replication. Thus, it was found that certain strains of bacteria which were allowed to complete their normal DNA replication cycles in the absence of protein synthesis were much more resistant to UV than those in exponential growth. However, protein synthesis inhibition in a UV-sensitive strain (E. coli $B_{s-1}$) had no appreciable effect on the UV survival curve for this strain (Figure 6-5). The excision-repair system has once again been implicated since this strain does not excise thymine dimers or perform repair replication.

It is known that bacteria that have completed their DNA replication cycles in the absence of required amino acids exhibit a lag in the resumption of DNA synthesis when the amino acids are re-added to the medium. This lag may provide the time necessary for optimum repair of damage before the cell again attempts semiconservative replication. Stationary phase cultures show a similar lag in resumption of normal DNA replication. It is of interest that nearly identical UV survival curves have been found for cultures of strain E. coli TAU that have been starved for amino acids and those that have been irradiated in early stationary phase.

Since repair replication can occur in the absence of protein synthesis it might be predicted that the maximum survival of bacteria for a given UV dose would be found if the culture were allowed to complete the normal DNA replication cycle prior to irradiation and then was not permitted to reinitiate the cycle until all possible repair synthesis had been completed. For a culture of E. coli TAU-bar in the state of completed DNA replication, the initial nonzero slope of the survival curve may reflect inactivation by nonrepairable damage (Figure 6-5A). The increased inactivation rate at high doses may be the result of the saturation of the repair replication system or the eventual inactivation of the repair system itself. In any case it is clear that the shape of the survival curve does not relate to a multiplicity effect (Section 6-4). To distinguish between saturation of the repair system and inactivation of the repair enzymes at high doses, it would be useful to compare the effects of 260 nm and 280 nm irradiation on repair replication.

### D.  Indirect Photoreactivation

The bacterial mutant that lacks the photoreactivating enzyme (Section 7-3,C2) does exhibit a photoreactivation effect when illuminated at 334 nm after UV irradiation. This effect is found to be temperature independent and it does not saturate at high intensities of illumination, suggesting that the process is nonenzymatic. The action spectrum and the dose requirement closely parallel that of photoprotection (Section 7–6,E) rather than that of enzymatic photoreactivation. Furthermore, it has been shown that this photoreactivation effect does not involve the splitting of thymine dimers. In studies in which the temperature and light intensity were varied during postirradiation illumination at 334 nm it was shown (using *E. coli* B) that at this wavelength the indirect photoreactivation mode constitutes an appreciable fraction of the total photoreactivation effect. The action spectrum for enzymatic photoreactivation exhibits a considerably lower efficiency at 334 nm than at 405 nm (Figure 7–4).

It has been hypothesized that the mechanism of the 334 nm effect is a temporary inhibition of growth and division. This hypothesis is supported by the finding that the action spectrum for growth inhibition in *E. coli* B closely follows the action spectrum for indirect photoreactivation. Thus, the proposed mechanism involves an enhancement of the effectiveness of the excision-repair system by delaying normal growth processes until repair is complete. However, it has not been shown that this treatment specifically inhibits DNA replication in bacteria and the detailed mechanism remains obscure. Since 334 nm radiation does reduce the quinone content of bacteria and since the addition of quinones to extracts from bacteria irradiated at 334 nm does restore electron transport activity, it has been suggested that the destruction of quinones could account for the phenomenon.

### E.  Photoprotection

A phenomenon quite similar to indirect photoreactivation in most respects is that of photoprotection (PP). Survival to UV irradiation is greater if bacteria have been subjected to 334 nm illumination *prior* to irradiation. The phenomenon has been shown in a number of strains of bacteria as well as in protozoa. Although photoprotection was observed in *Pseudomonas,* it was not seen in a photosynthetic bacterium of the same family. The somewhat spotty occurrence of PP may be due to the fact that radiation in this wavelength region (310 nm to 370 nm) has an additional inactivating effect on some organisms and this may obscure the observation of recovery. The explanation for PP, as well as for the indirect photoreactivation mechanism, most probably relates to growth inhibition effects that allow more time for repair of damage in DNA (Figure 7-8).

## F. UV Reactivation

The UV survival of λ bacteriophage was shown to be enhanced if the host bacteria were given a low dose of UV radiation prior to infection. In fact, some enhanced survival could be observed if the λ-host complex was irradiated lightly with UV after infection. The mechanism of action of the process is quite confusing since it would appear to be contradictory to the process of host-cell reac-

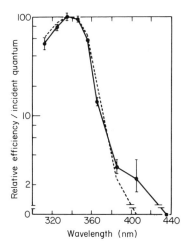

FIGURE 7-8. An action spectrum for growth delay and for photoprotection in *E. coli* B. —, Growth delay in nutrient broth; ---, photoprotection. [From J. Jagger, W. C. Wise, and R. S. Stafford, *Photochem. Photobiol.* 3, 11 (1964).]

tivation, even though it generally is found in systems that exhibit HCR. The fact that UV reactivation (UVR) can be reduced by photoreactivation indicates that the effect is on DNA. Further evidence for this site of action is the finding that other agents that produce repairable damage in DNA, also result in reactivation effects similar to UVR. Thus, one is faced with the apparent situation that damage to the host genome may enhance the repairability of the damaged phage genome even though the effect expected would be one of competition of the two damaged genomes for available repair enzymes. Very little UVR is observed in bacteria that are deficient in excision-repair.

## G.  Filament Formation and Recovery of Cell Division

In this section we have lumped together a series of poorly understood phenomena which are classically documented by the differences between the E. coli strains B and B/r. These strains are able to perform excision-repair with similar efficiencies but they exhibit strikingly different survival characteristics. Cytological differences are seen between UV-irradiated cultures of E. coli B and B/r. Strain B shows inhibition of cell division even at such low UV doses that no effect on DNA replication is detectable. Thus, the cells continue to increase in mass and grow into long filamentous forms eventually reaching lengths 50 times that of normal cells. However, many of these filaments may eventually recover division ability under appropriate conditions and may then give rise to viable colonies upon agar plates. Upon reirradiation these filamentous cells are much more UV sensitive than nonfilamentous cells. Mutants of E. coli K-12 have been isolated which also manifest this extreme sensitivity to UV inhibition of division. The phenomenon of recovery from division inhibition in these strains has been called K-reactivation by Kneser. However, the mechanism of this repair system (if it is indeed a repair system) remains obscure. These division inhibition phenomena seem to involve only lethal lesions since UV mutagenesis, macromolecular synthesis, prophage induction, phage inactivation, etc., are not affected by the mutations that promote filament formation.

The UV-induced inhibition of division in strain B and K-12 LON⁻ (K-reactivation deficient) can be reversed by incubation of the cells at 45°C after irradiation, by the addition of pantoyl lactone to the growth medium, or by the incubation of the cells in liquid medium at high culture densities. This latter effect has been termed "neighbor restoration." Survival of *E. coli* B was found to be a function of the density of bacterial cells on the plating agar, even if cells of another strain contributed to this density. In this connection it is interesting that a "division promoting" substance believed to be enzymatic has been isolated from bacterial extracts. The recovery of division ability produced by 45°C incubation might account in part for the phenomenon of thermal reactivation (Section 7-3,A).

It is clear that an adequate sorting out of the different types of reactivating effects may require an understanding of the genetic markers that are involved and how they are related. This will be considered briefly in the final section below.

## 7-7. GENETIC CONTROL OF REPAIR PROCESSES

The isolation of UV-resistant and UV-sensitive mutants of bacteria provided one of the strongest pieces of indirect evidence for the existence of repair processes in cells. The first radioresistant mutant was the strain B/r isolated from *E. coli* B by Evelyn Witkin and the first sensitive strain $B_{s-1}$ was isolated from *E. coli* B by Ruth Hill. The strain $B_{s-1}$ was shown to have an abnormally low plating efficiency for UV-irradiated phage T1 and it was suspected that it therefore lacked some factor involved in the reactivation of photoproducts in the phage. Similar mutants were obtained from *E. coli* K-12 by Howard-Flanders and Theriot. Bacterial conjugation was used by Howard-Flanders and co-workers to identify the locus responsible for UV sensitivity on the genetic map, and it was found that a locus situated between the arginine and arabinose markers on the male chromosome could confer UV resistance to the progeny of the zygotes. It was suggested that the UV-resistance locus controlled an enzyme system capable of reactivating in the dark the same UV photoproducts that were photoreactivable in the light. Furthermore, the finding that the replacement of thymine by 5BU in T1 phage inhibited reactivation led to the suggestion that the action was on thymine photoproducts in DNA. Many more UV-sensitive mutants have now been isolated and the specific biochemical defects involved in the various repair systems are beginning to be understood. It should be emphasized that it may be difficult to isolate certain types of mutants deficient in repair processes since some of the steps in these processes may also be necessary for normal cell growth.

The known mutant loci that are involved in repair are shown on a genetic linkage map of the *E. coli* chromosome in Figure 7-9 along with some other genetic markers for reference. The terminology for these different markers can be somewhat confusing since it has arisen in a number of different laboratories simultaneously. Thus, for example the DIR, FIL, and LON designations all refer to the same phenomenon, the sensititity of division ability to radiation. Rörsch has suggested that a convention be established in which the mutant phenotypes (e.g., DIR) be designated by capital letters and the genotypes be indicated by lower case letters.

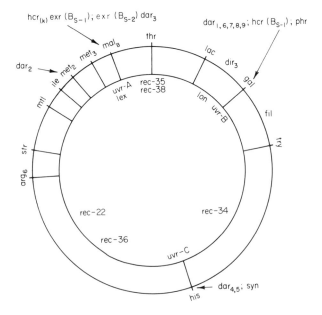

FIGURE 7-9. A linkage map of the *E. coli* chromosome, showing the positions of various markers known to affect radiation sensitivity. Most of these are also known to affect specific repair mechanisms. [Adapted from P. C. Hanawalt, *in* "Photophysiology" (A. C. Giese, ed.), Vol. 4, p. 244. Academic Press, New York, 1968.]

A number of different classes of radiation sensitive mutants have been designated as follows:

(a) FIL, DIR, or LON: Mutants that tend to stop dividing after low doses of UV. Filament formers. (Section 7-6,G) Genetic loci: lon, dir, fil.

(b) HCR: Mutants that are deficient in host cell reactivation and are also sensitive to UV. Some of these have been shown to be specifically deficient in dimer excision. (Section 7-4,B) Genetic loci: hcr, dar, $dar_3$, $dar_5$, $dar_6$, uvrA, uvrB, and uvrC, syn. UVR is used to designate the phenotype of the uvrA, uvrB and uvrC genotypes. This can cause some confusion in view of the next phenotype class designated by Rörsch as UVR.

(c) UVR: mutants that are UV sensitive but *not* deficient in host cell reactivation of infecting phage. These mutants are generally not unusually sensitive to X-rays. Genetic loci: $dar_2$, $dar_4$, $dar_7$, $dar_8$, $dar_9$, $B_{8-2}$.

(d) EXR: Mutants that are sensitive both to UV and to X-rays. Genetic loci: exr.

(e) REC: Mutants that are UV sensitive, X-ray sensitive, and deficient in genetic recombination. Genetic loci: rec.

(f) PHR: Mutants that are deficient in enzymatic photoreactivation. Genetic locus: phr.

Most viruses must depend upon the repair facilities available in the host cell for dealing with photochemical (and other) damage in their own genomes. However, the bacteriophages T2, T4, and T6 have been shown to contain genetic information for the repair of photochemical damage. One of these processes (discovered by Luria) is controlled by the v-gene in phage T4 and is not normally present in T2 or T6. Harm found that heavily UV-irradiated T4 could, upon mixed infection of cells with UV-irradiated T2, contribute to the enhanced survival of the T2 phage, thus indicating that reactivation was due to the $v^+$ allele. The complete independence of v-gene reactivation and photoreactivation was demonstrated with a photoreactivationless bacterial mutant. The mechanism of v-gene reactivation has been shown to involve the excision of thymine dimers from the damaged phage DNA.

A second gene, present in T2, T4, and T6, was discovered by Harm and termed the x-gene. The x-gene also apparently controls an intracellular repair mechanism. The initial photoreactivable

UV damage to all four possible combinations of v and x genes in T4 was found to be identical. Also, the allelic state of the x-gene in unirradiated parental phage was found to determine the extent of marker rescue from an irradiated parental phage. This has led to the suggestion that the x-gene may affect the process of genetic recombination in the phage.

In conclusion, the importance of gene controlled enzymatic mechanisms for the repair of photochemical damage in biological systems is emphasized by the fact that genomes as small as those of the T-even bacteriophage contain information for such mechanisms. It is in this area of the genetic control of cellular repair processes that perhaps the most exciting discoveries will be made in the next few years.

## GENERAL REFERENCES

C. S. Rupert, Photoreactivation of ultraviolet damage. *In* "Photophysiology" (A. C. Giese, ed.), Vol. 2, p. 283. Academic Press, New York, 1964.

J. K. Setlow, Photoreactivation. *Radiation Res.* Suppl. 6, 141 (1966).

R. B. Setlow, Repair of DNA. *In* "Regulation of Nucleic Acid and Protein Biosynthesis" (V. V. Koningsberger and L. Bosh, eds.), p. 51. Elsevier, Amsterdam, 1967.

P. C. Hanawalt, Cellular recovery from photochemical damage. *In* "Photophysiology" (A. C. Giese, ed.), Vol. 4, p. 204. Academic Press, New York, 1968.

P. Howard-Flanders, DNA repair. *Ann. Rev. Biochem.* 37, 175 (1968).

R. B. Setlow, The photochemistry, photobiology, and repair of polynucleotides. *Progr. Nucleic Acid Res. Mol. Biol.* 8, 257 (1968).

B. S. Strauss, DNA repair mechanisms and their relation to mutation and recombination. *In* "Current Topics in Microbiology and Immunology," Springer, Berlin, 44, 1 (1968).

# 8

# Ultraviolet Mutagenesis

## 8-1.  INTRODUCTION

Following the discovery that ultraviolet light kills bacteria, it was found that UV is a very efficient mutagen in the cells that survive. For low UV doses the mutation frequency may be increased $10^3$- to $10^6$-fold over the spontaneous mutation level and for high UV doses nearly all of the survivors may be mutants. A natural question then to ask is to what extent does the observed lethality depend upon the production of lethal mutations? The answer implied in earlier chapters and to be elaborated here is that lethal mutations really play a very minor role in the UV inactivation of cells. Ultraviolet damage to DNA may be broadly classified in two categories which are probably not mutually exclusive: some lesions may impair replication while others may permit replication but with a high error frequency. The study of UV mutagenic mechanisms is a very active current area of research but, in addition, UV is used simply as a method for obtaining mutants. Ultraviolet radiation has probably been the most widely used mutagen for the routine production of microbial mutants for genetic studies. More recently, chemical mutagens (e.g., nitrosoguanidine) have been discovered which provide still higher ratios of mutants to survivors. The ideal reagent

165

for efficient mutant production should be one that produces changes in DNA that are not recognized by repair systems. This may be true for nitrosoguanidine but is is not generally true for UV. The relationships between UV damage to DNA, its repair, and mutagenesis are very complex, as we shall see.

The action spectrum for UV mutagenesis (See Figure 1-11) was shown to follow a nucleic acid absorption spectrum. Thus, our consideration of the UV induction of mutants should follow the general pattern of our previous consideration of the lethal effects of UV photons. Lesions are produced in DNA and these presumably can lead to errors in the incorporation of the proper nucleotides into growing polynucleotide strands during subsequent replication and transcription. The repair of these lesions by one or the other of the mechanisms discussed in the previous chapter may prevent the expression of the mutation and may restore the DNA to its normal undamaged state. On the other hand it must be realized that the repair system itself might also produce mistakes. In fact, a lesion in DNA that would not otherwise result in a mutation could conceivably stimulate the repair of the damaged region by a process that would leave a mutation in its place. Finally, a possibility to be considered is that UV may damage some of the precursors for DNA synthesis and that the subsequent incorporation of such damaged precursors (base analogs in effect) might lead to mutation. Early observations in a number of laboratories indicated that UV irradiation of the substrates for bacterial growth could lead to increased mutagenic frequencies in bacterial cultures. However, the mutation yields obtained were always much lower than those produced by direct irradiation of the bacteria. Also, the most efficient wavelengths for producing mutagenic substrates were those below 2000 Å. Thus, the contribution of such effects to UV mutagenesis is probably negligible, especially in experiments employing filtered germicidal UV lamps (2537 Å).

There are several basic classes of mutations to be considered; these are classified with respect to the type of alteration produced in DNA. The *transition* mutations are those in which a pyrimidine is replaced by a different pyrimidine (e.g., deoxycytosine by deoxyuridine or thymidine) or a purine is replaced by a different purine. At the next replication of the altered region a similar transition will be induced in the daughter strand (i.e., pyrimidine transition will result in a purine transition in the newly synthesized complementary strand). This will be a consequence of the hydrogen bonding properties of the bases involved. However, even a mispaired pyrimidine would probably not appreciably distort the backbone structure of the DNA and it is not very likely that the excision-repair system would detect such a minor alteration. On the other hand a *transversion* mutation, in which a pyrimidine is replaced by a purine (or vice versa), would necessarily result in a distortion of the DNA backbone in addition to base pairing difficulties, since two purines hydrogen-bonded in the helix would increase the diameter of the molecule while two hydrogen, bonded pyrimidines would decrease it. It is possible that the excision-repair system might recognize this sort of distortion of the DNA backbone structure.

A third type of mutation has the effect of adding or deleting a nucleotide (or nucleotides) and this may cause a *reading frame shift* (or sign mutation) such that the codon triplets following the damaged region are no longer in register and incorrect amino acids would be incorporated into the resultant proteins. Sooner or later the message would contain a nonsense codon (i.e., a triplet of nucleotides that does not code for an amino acid) and reading would be terminated. This class of mutation may be due to the actual removal of one or more nucleotides from a template strand or it may be induced by the intercalation of a molecule (e.g., an acridine dye such as proflavin or acriflavin) between two nucleotides in a strand of DNA. The intercalation of a planar dye molecule between two stacked planar bases in a polynucleotide strand distorts the structure in such a way that a third nucleotide is inserted in the complementary strand in the course of the replication or transcription of the damaged strand. That is, the nucleotide plus dye plus nucleotide combination is read as a triplet of nucleotides and, as a consequence, subsequent triplets would be read "out of phase" by one nucleotide. It may be difficult to distinguish between an actual deletion and a molecular distortion that results in a deletion (or addition) when the damaged strand is replicated.

Another theme to be developed is the relation of the physical state of the DNA and its activity (with regard to transcription and replication) to the production and subsequent expression of mutagenic damage. Herein may lie the key to an understanding of the molecular basis of UV mutagenesis. As we have seen in previous chapters the relative yields of different photoproducts in DNA are dependent in part upon the state of denaturation of the DNA. Some photoproducts (e.g., hydrates) are much more readily produced in denatured DNA than in native DNA. An unknown factor that may also be important is the binding of other molecules to the DNA. For example, a region of DNA that is temporarily in association with an RNA polymerase might exhibit unique photochemical responses. Regions of the genome that are being transcribed much of the time may exhibit different sensitivities to mutation than regions that are seldom transcribed.

## 8-2. EARLY OBSERVATIONS ON BACTERIAL SYSTEMS

Just as the survival of irradiated cells is often dependent upon the treatment of the culture before the plating assay for viability, the mutagenic response of cells also depends upon how soon after irradiation one scores for the mutations. It was shown in an early study of UV mutagenesis in *E. coli* B/r that two distinct categories of mutants could be distinguished and these showed different dependencies upon dose. The UV dose-effect curves for the mutation from phage T1 sensitivity to T1 resistance of *E. coli* B/r is illustrated in Figure 8-1. The two categories are the so called *zero point* mutations and the *end point* mutations. Zero point mutations are those that are expressed immediately: it can be seen that the number of mutations in this category increases with dose at a rate that exceeds an exponential function of dose, but then a plateau occurs followed by a decline at very high doses where survival is less than 0.1%. In contrast, the end point mutations that are scored after a period of growth exhibit a rapid increase at low doses and then a more gradual increase with no apparent maximum. Note that end point mutations are always in considerable excess over zero point mutations and in fact the scoring of the end point mutations includes the zero point mutations. A pattern quite similar to that obtained with UV is seen when X-rays are used as the mutagenic agent.

The situation for mutagenesis to T1 resistance in *E. coli* appears quite complex. However, before drawing any general conclusions one must consider the response of other alleles to UV. A linear increase with UV dose was obtained for the reverse mutation of a streptomycin-dependent strain of *E. coli* B/r to streptomycin inde-

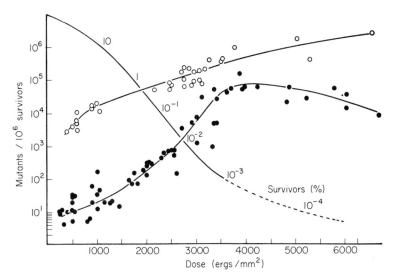

FIGURE 8-1.  Zero-point [●] and end-point mutations [○] to T1 resistance in *E. coli* B/r, as induced by 254 nm UV radiation. [Adapted from M. Demerec and R. Latarjet, *Cold Spring Harbor Symp. Quant. Biol.* 11, 38 (1946).]

pendence. In this case no zero-point mutations would be observed since this class of mutation is not expressed until one division has occurred. A quite different response to UV mutagenesis was seen in studies on color-response mutants of *E. coli* B/r on mannitol tetrazolium agar. The mutation frequency rose very rapidly below about 500 ergs/mm². No decline in mutation frequency occurred at higher doses. In a study of UV-induced reverse mutation in different *E. coli* mutants Demerec and co-workers found that the spontaneous rate of back mutation could not be increased at all in 4 of the 27 strains tested. Thus, it was clear from early studies that not all bacterial mutations exhibit the same sensitivity to UV and that some may not respond to UV at all.

In Chapter 6 we saw that *E. coli* strains B and B/r exhibited different UV inactivation rates. However, both strains were found to yield the same rates of mutation to T1 resistance. This finding led to an early suggestion that lethal mutation might be of minimal importance in the killing of bacteria. Obviously if the difference between the two strains was simply the efficiency of repair and if the repair was mostly in response to mutation damage, then B/r should have presented less mutability as well as higher survival than strain B. Of course, it is now known that the difference between these two strains is in a characteristic not directly related to excision-repair (Section 7-6).

A classic paper was presented in 1956 at the Cold Spring Harbor Symposium by Evelyn Witkin, describing her efforts to obtain more detailed information on the timing of the mutation process. The studies were carried out with amino acid auxotrophs of *Salmonella typhimurium*. Induced prototrophs could be obtained from these auxotrophs following UV irradiation of the bacteria. The mutation to prototrophy could also be transferred to an auxotroph by infection with transducing bacteriophage that had been grown on a stock of UV-induced prototrophs. A transducing phage is one that has accidentally incorporated some segments of the bacterial genome in the course of induction of the prophage. A comparison was made between the pattern of the delayed appearance of UV-induced mutants and the pattern of appearance of similar mutants arising by transduction. In the latter case the genetic change is brought about by a recombinational event and no mutagenic action is involved. This sort of comparison was designed to determine whether a phenotypic lag was involved in the mutation delay. The maximal yield of prototrophs arising by transduction was found under conditions permitting only one residual division of the cells. The maximal yield of UV-induced prototrophs was seen on agar plates enriched with nutrient broth to permit nearly six residual divisions. The results showed clearly that the delayed appearance of induced prototrophs could not be explained by phenotypic lag since this lag did not exceed about one division period. However, as Witkin pointed out, the finding that the phenotypic lag for induced prototrophs was less than one generation did not exclude the possibility that other classes of mutations might exhibit more prolonged lags. Thus, a ten to twelve generation delay was observed for the optimal expression of a mutation to phage resistance.

The total yield of induced prototrophs was shown to depend upon the amount of nutrient broth present in the culture medium during the first hour of postirradiation growth—the active component for this effect was found to be the amino acids. The presence of chloramphenicol to inhibit protein synthesis during the first hour after irradiation was shown to drastically reduce the mutant yield; consistent with a requirement for protein synthesis for the optimum yield of mutants. This important observation provided inspiration for similar studies in a number of laboratories and a ten-year period of elaboration without elucidation.

The temperature of postirridiation incubation was also shown by Witkin to have a profound effect upon mutant yield while having much less effect upon survival level. The yield of induced prototrophs was increased 5-fold over that obtained at 37°C if short incubations at 37°C were followed by incubation at lower temperatures. On the basis of these temperature effects and the known dependence of mutant yield upon the rate of protein synthesis Witkin postulated the involvement of a repair system in UV mutagenesis. It was supposed that protein synthesis favored the repair of UV-damaged DNA and that the survivors had a high probability of being mutants. This explanation would appear to be just the opposite of that proposed to explain the effect of liquid holding upon viability, and this situation is one of many instances in which viability and mutagenesis are not linked in a clear way.

Doudney and Hass reported an extension of Witkin's observations in a series of papers in which they also coined some new terminology to describe the time course of mutant yield under different conditions of incubation after irradiation. The phenomena are illustrated in Figure 8-2. Under conditions of protein synthesis inhibition the yield of UV-induced reversions of an *E. coli* tryptophan auxotroph was found to decrease with time and this was called *mutation frequency decline*. If protein synthesis was enhanced the yield of mutants was seen to increase for about one generation period and then to remain constant: this was termed *mutation stabilization*. The presence of dinitrophenol to uncouple oxidative phosphorylation resulted in no significant increase or decrease in the mutation frequency. The addition of amino acids to a culture in the course of the mutation frequency decline did not increase the mutant frequency but caused it to remain constant at the level attained when the amino acids were added. Once mutation stabilization had occurred the inhibition of protein synthesis had no further effect upon mutant yield.

In subsequent studies Doudney and Hass showed that several

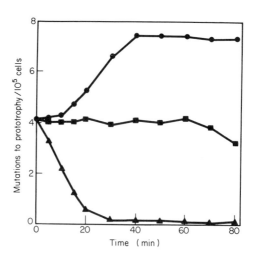

FIGURE 8-2. Comparison of "mutation stabilization" and "mutation frequency decline" in the UV-induced reversion of a tryptophan auxotroph of *E. coli* to prototrophy. The irradiated cells were incubated for the times shown in the different media before plating on nutrient broth supplemented agar plates. After 3 days incubation the plates were scored for mutation and survival. No significant changes in survival level were produced by the treatments indicated. ●, Minimal medium plus added casein hydrolyzate; ■, minimal medium plus added casein hydrolyzate plus dinitrophenol; ▲, minimal medium plus added casein hydrolysate plus chloramphenicol. [Redrawn from Figure 1 in C. O. Doudney and F. L. Haas, *Proc. Natl. Acad. Sci. U.S.* **45**, 709 (1959).]

antimetabolites that inhibit RNA synthesis also prevented mutation stabilization. However, Leib found that the increases in mutation stabilization levels more closely followed DNA synthesis than RNA synthesis. The experiments of Weatherwax and Landman with a thymine-requiring bacterial strain showed that mutation stabilization was delayed if thymine was withheld. Thus, they concluded that mutation stabilization was established by DNA synthesis. Their model to explain these observations implicated repair processes in mutation frequency decline. It was supposed that the possibility of repairing the damage is terminated when the damaged region is passed by the normal replication point. More recent evidence suggests that damage can be repaired even after replication. An altered nucleotide sequence in the newly replicated daughter DNA strands, however, would be indistinguishable from unaltered DNA as far as the recognition step in repair is concerned.

Now it is important to realize that the studies which led to the above pattern of mutation expression were carried out with reversions of auxotrophs to prototrophy. Not all types of mutations require postirradiation protein synthesis for stabilization. Mutations to streptomycin independence were shown by Witkin and Theil not to respond to the various manipulations of protein synthesis following UV irradiation. Thus, it would seem that some peculiarity of mutations to prototrophy is responsible for the marked dependence upon postirradiation protein synthesis and/or DNA synthesis. We will return to this problem after a consideration of the chemical nature of UV mutagenesis.

## 8-3. NATURE OF THE MUTANTS PRODUCED BY UV

In Section 4-4 we considered one approach to the chemical nature of UV mutagenesis. Synthetic polynucleotides of known composition are irradiated with UV and the resultant material is used as a template for *in vitro* RNA (or protein synthesis). The chemical nature of the RNA (or protein) product is examined and any abnormalities may provide information on the molecular nature of the mutation. Thus, irradiated poly-C was found to lose its ability to stimulate the incorporation of guanosine triphosphate (GTP) in an *in vitro* RNA polymerase system but adenosine triphosphate (ATP) was then incorporated. The effect was heat reversible. From this one might imagine that a mutagenic action of UV on DNA could involve the formation of a cytosine hydrate.

Another approach to the chemical nature of mutations involves the application of various chemical mutagens of known specificity in attempts to revert the UV-induced mutations. The determination of reversion characteristics can then be used to infer the nature of the original mutation. Using the phage T4 system Drake obtained several hundred mutants in the rII region of the genome by UV irradiation of infected cells. An advantage of using the rII system is that although such mutants do not grow on K strains of *E. coli* they all grow on B strains. Thus, it is possible to recover all of the mutants in viable phage. This is of obvious importance to any meaningful characterization of the spectrum of mutants produced. The mutants were mapped and one member at each locus was then examined for revertability by base analogs and by proflavin. Base analogs could not revert 50% of the mutants. Of these, 36 out of 39 were reverted by proflavin, which produces reading frame shift mutants. None of the base analog revertable mutants could be reverted by proflavin. However, the fact that some mutants could be reverted by this mutagen indicates that a sizable fraction of the UV-induced mutants were reading frame shift mutants. This sort of mutation would be expected if it were possible for the DNA polymerase to synthesize DNA around a pyrimidine dimer and occasionally add an extra nucleotide in the process.

The mutants in the T4 system that were revertable by base analogs (i.e., half of them) were broadly classed as transition mutations. These were further subdivided into those which responded to reversion by hydroxylamine and those that did not. Hydroxylamine specifically induces transitions such that a GC-base pair will lead to a TA-base pair upon replication. It turned out that none of the mutants were revertable by hydroxylamine. Thus, the GC behaved as an AT pair and, as discussed above, this is the same specificity expected for the hydration product of cytosine.

Since photoreactivation is known to act specifically upon UV-induced pyrimidine dimers it was of interest to examine the effect of photoreactivation upon the mutants produced in this system. Oddly enough, no difference in mutant frequency was seen after photoreactivation.

In a similar study, Tessman examined the specificity of UV mutagenesis in phage S13 containing single-stranded DNA. This system was a bit peculiar in that no mutants were obtained if only the phage were irradiated. If both the phage and the host bacterium were irradiated separately prior to infection the survival level of the phage was higher and mutants were obtained. This phenomenon of UV reactivation has been discussed in Section 7-6,F. Of 16 UV-induced mutants in the S13 phage system 11 were shown to be transition mutants of the cytosine to thymine type. The other 5 were not clearly characterized but could be transition mutants of the cytosine to thymine type or else they could be sign mutants. Again no revertants of UV mutants were obtained by the action of hydroxylamine. These results are consistent with those of Drake on the T4 phage system and they also implicate cytosine as the altered base. Note, however, that Drake had to infer that the target was cytosine rather than its complementary guanine whereas the involvement of cytosine was unambiguous in the single-stranded DNA of phage S13.

It is important to realize that in both these phage systems the host was irradiated as well as the phage. Thus, one can worry that the spectrum of mutants produced is in some way dependent upon the irradiation of the host. Indeed, no mutants were produced unless the host was irradiated prior to infection by the S13 phage. It is not known why irradiation of the host influences phage mutability. One might suppose that the UV destroys some natural capacity of the host for reversing mutagenic damage in the phage.

An elegant method for studying the molecular specificity of mutagenesis involves the examination of the specific defect in a protein synthesized *in vivo* in response to the mutant gene. This method has been utilized by Yanofsky and co-workers to study the tryptophan synthetase enzyme system in *E. coli*. In some of the mutants the location of the defect has been shown to involve the insertion of an incorrect amino acid in the polypeptide chain. Thus, for example, one such replacement (of a glycine by an arginine) was found to be caused by the substitution of two guanine residues for an adenine-cytosine sequence in the DNA. A similar approach has been used with several other proteins of known amino acid sequence.

Some UV-induced mutations have been found to be revertable by UV irradiation. If the molecular type of mutation were quite specific for UV then one might wonder how such reversions could be possible. Thus, if UV produces a C to T transition, how can subsequent UV irradiation cause a T to C transition? The answer is that it probably doesn't work that way. A reversion could be due to an alteration (i.e., another amino acid replacement) elsewhere in the protein that restores the protein to functionality. Also, the primary mutation may be suppressed as a result of a mutation at some site other than in the operon in question. This type of suppression can result in a genetic alteration in the specificity of amino acid incorporation in response to a given codon. Such suppressor mutations would be expected to occur specifically in regions of the genome that are involved in the translation of the code (i.e., in the regions that specify the transfer RNA species and perhaps those that determine the ribosomal components, RNA and protein). The importance of suppressor mutations in UV mutagenesis is highlighted by the fact that those revertants to prototrophy that exhibited mutation frequency decline in the absence of protein synthesis have turned out to be suppressor mutations.


## 8-4. ROLE OF REPAIR PROCESSES IN MUTAGENESIS

The chemical nature of the UV photoproducts responsible for mutagenic effects can sometimes be inferred from the actions of repair mechanisms of known specificity. As discussed in Section 7-3, one such repair mechanism, enzymatic photoreactivation, is very specific and has been shown to deal only with UV-induced pyrimidine dimers. Thus Novick and Szilard found that certain mutations were about as much affected by photoreactivation as was lethality in UV-irradiated bacterial cultures. Similar results were reported for other biological systems such as *Paramecia, Neurospora,* and *Drosophila.* An examination of such results in the light of current knowledge of molecular mechanisms would lead one to believe that UV mutations are mostly due to pyrimidine dimers. If these dimers are repaired then mutations do not occur. Unfortunately, the situation is much more complex in that many factors are interrelated in the determination of mutation probabilities. Even in early studies on the photoreactivation of UV mutation Newcombe and Whitehead found that photoreactivation after low doses of UV resulted in a dose-reduction factor of 5 (greater than that for bactericidal effects) while for large doses no photoreactivation effect was noted. Also, photoreactivating light has no effect on the production of UV-induced lac$^-$ mutations in the excision deficient *E. coli* strain WP2$_s$ of Witkin.

The study of this same strain has supported the involvement of pyrimidine dimers in UV mutagenesis, however. This strain has been shown to revert to prototrophy (tryptophan independence) by either of two mechanisms: a base substitution at a nonsense triplet in the tryptophan locus, or by the induction of suppressors. These suppressors have been shown to be mostly base changes in the gene that determines glutamine transfer RNA. It has been shown that over 90% of the UV-induced mutations in this strain are reversed by photoreactivation and this seems to be true for both those mutants localized at the nonsense triplet and for the suppressor mutations. Bridges and Munson studied the production of mutants in this strain following growth in the dark after low doses of UV. The fraction of photoreactivable prototrophic mutations was assayed after different periods of postirradiation growth. The results indicated that dimers at mutable sites persisted for nearly 4 generation periods after the irradiation, and that they could still give rise to mutations with a low probability at each replication cycle during this period.

Some results are not explained adequately by the presence or absence of functional repair systems. Among them are those mutants that respond to mutation frequency decline. The principal clue is that these mutations invariably turn out to be of the suppressor type. It has been suggested by Witkin that protein synthesis (that leads to mutation frequency stabilization) may be causing a change in the physical state of the DNA at the suppressor loci such that the repairability of UV lesions in these regions is reduced. It might be expected that the excision repair system would not be able to operate effectively in a region in which the DNA strands were separated since an intact complementary DNA strand is needed to provide the specificity for repair synthesis. The genes that code for transfer RNA and the various ribosomal components are exceptionally active during conditions of normal growth. Thus, under such conditions, UV damage in these regions might be less repairable. The inhibition of protein synthesis would reduce the activity in these regions and the DNA would then exist in the double-stranded repairable state. Therefore, the suppressor mutations could be repaired during a period of protein synthesis inhibition and mutation frequency decline would be observed.

In connection with the above general model another possibility for mutagenesis could be envisaged in which repair of UV-induced damage would occur on the transient RNA–DNA hybrid that exists at a moment of transcription of a suppressor locus (or any other locus for that matter). The RNA strand might be used as a template for the "patching" step and the transcription of the RNA region opposite the original lesion could lead to errors which would be incorporated into the DNA strand in subsequent repair. Much more research is required to implicate such models.

Finally, the unique properties of the EXR⁻ bacterial strains should be considered. These strains (Section 7-7) are characterized by the property that they are sensitive to UV and X-rays but that they are still able to promote the host cell reactivation of infecting phage. It is a curious property of these strains that no mutants can be produced by UV. To explain this phenomenon Witkin postulates the presence of a less efficient, but more accurate system for repairing mutation-producing lesions in exr⁻ strains than in exr⁺ strains. The detailed study of UV mutagenesis may result in the discovery of additional mechanisms for dealing with photochemical damage in DNA.

## GENERAL REFERENCES

M. R. Zelle and A. Hollaender, Effects of radiation on bacteria. *In* "Radiation Biology" (A. Hollaender, ed.). Vol. 2, p. 365. McGraw-Hill, New York, 1955.

E. M. Witkin, Time, temperature and protein synthesis: A study of ultraviolet induced mutation in bacteria. *Cold Spring Harbor Symp. Quant. Biol.* 21, 123 (1956).

G. Zetterberg, Mutagenic effects of ultraviolet and visible light. *In* "Photophysiology" (A. C. Giese, ed.), Vol. 2, p. 247. Academic Press, New York, 1964.

J. W. Drake, Studies on the induction of mutations in the bacteriophage T4 by UV irradiation and by proflavin. *J. Cellular Comp. Physiol.* 64, Suppl. 1, 19 (1964).

B. D. Howard and I. Tessman, Identification of the altered bases in mutated single-stranded DNA: III. Mutagenesis by ultraviolet light. *J. Mol. Biol.* 9, 372 (1964).

E. M. Witkin, Radiation induced mutations and their repair. *Science* 152, 1345 (1966).

E. M. Witkin, Mutation-proof and mutation-prone modes of survival in derivatives of *E. coli* B differing in sensitivity to ultraviolet light. *Brookhaven Symp. Biol.* 20, 17 (1968).

C. O. Doudney and F. L. Hass, Modification of UV-induced mutation frequency and survival in bacteria by post-irradiation treatment. *Proc. Natl. Acad. Sci. U.S.* 44, 390 (1958).

# 9

# Photodynamic Action

## 9-1. INTRODUCTION

Thus far we have only been concerned with the direct interaction of photons with important biological molecules. We now turn to the situation where the energy of a photon is absorbed by a second molecule (a photosensitizer). The excited photosensitizer can then transfer its energy by various mechanisms to the target molecule. We have previously examined one case where visible light plus a sensitizer can be used to form thymine dimers, a product which can usually only be formed by the direct absorption of ultraviolet light by thymine molecules (Section 4-5).

The case we wish to examine now is *photodynamic action*, the oxidation of biologically important molecules in the presence of oxygen, a dye, and visible light. In this process oxygen and substrate are consumed but the dye is reused. Photodynamic action is thus rigidly defined. Dye sensitized photooxidations can also occur in the absence of oxygen (Section 9-7).

179

The discovery of photodynamic action is usually attributed to Raab who in 1900 observed that low concentrations of acridine orange (and other dyes) had no effect on *Paramecia* in the dark but rapidly killed them in the light. It was soon determined that oxygen was required for this reaction.

Many types of damage have been induced in bacterial systems by photodynamic action: (a) loss of colony-forming ability (death); (b) damage to DNA as evidenced by mutation production and base damage; (c) damage to cell membrane as evidenced by altered permeability; (d) damage to protein as evidenced by the inactivation of enzymes. In order to elucidate the mechanisms of photodynamic action on cells, many studies have been performed on the chemical and physical changes produced in nucleic acids and proteins by dyes plus visible light. In some cases it is known that certain dyes bring about the alteration of a specific nucleic acid base or a specific amino acid. With the increasing availability of specific reactions of this type, dye-sensitized photooxidation can be used as a selective tool to probe, for example, the nature of the amino acids in the active center of an enzyme. Some success with this approach has already been achieved (Section 9-6).

## 9-2. STRUCTURE OF PHOTODYNAMIC DYES

The structure of the majority of the compounds that are photodynamically active is based upon the anthracene nucleus, but the photodynamic effectiveness of these dyes is dependent in large measure upon the nature of the atoms in the meso portion of the ring (Figure 9-1). In order of increasing effectiveness the meso atoms are $(C,C) \leqq (C,O) \leqq (N,N) \leqq (N,O) \leqq (C,N) \leqq (N,S)$. Also a change from a planar arrangement of the rings to one in which some of the rings may be displaced out of the plane results in a marked decrease in photodynamic effectiveness. A bend in the ring

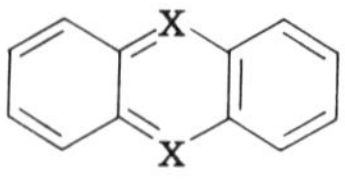

FIGURE 9-1.   Skeletal structure of many photodynamically active dyes. The X marks indicate the meso positions of the anthracene nucleus.

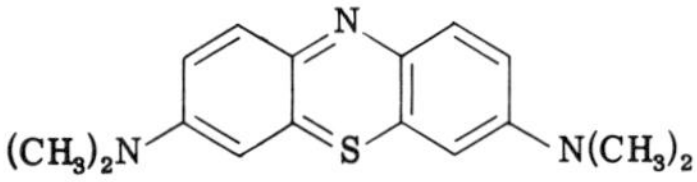

Methylene blue

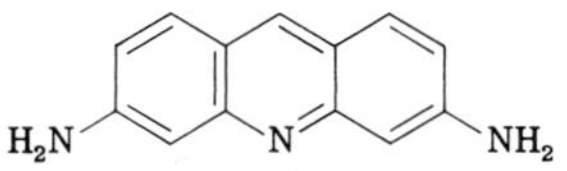

Proflavine

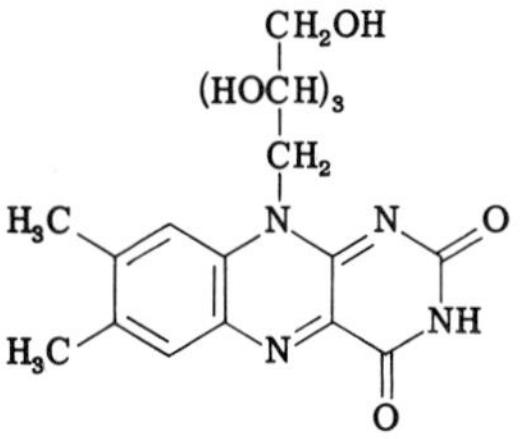

Riboflavin

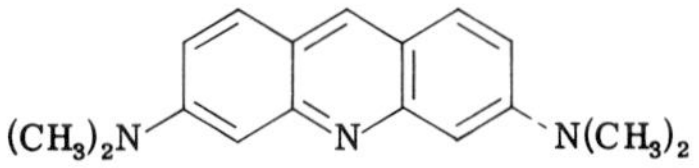

Acridine orange

FIGURE 9-2.  Structure of some photodynamically active dyes.

structure may sterically restrict the dye from coming optimally close to a target molecule. The structures of selected photodynamic dyes are given in Figure 9-2.

Since the light energy for producing a photodynamic effect must initially be absorbed by the dye, one can predict that the action spectrum for the photodynamic oxidation of a target molecule will mimic the absorption spectrum of the particular dye being used rather than that of the target molecule. In practice this means that light in the wavelength range of 3000–8000 Å is photodynamically active (the range of light absorption by most dyes).

A notable feature of dyes that produce photodynamic effects is that they fluoresce. The act of fluorescence is probably not important in the mechanism of photodynamic action (all fluorescent compounds are not photodynamically active) but it constitutes one form of evidence that the dyes have a molecular structure that is capable of excitation to metastable states by the absorption of radiation (see Chapter 3). Actually the excited state that gives rise to fluorescence (first excited singlet state) probably is not involved in the photodynamic process because it has too short a lifetime ($\sim 10^{-9}$ seconds) to allow for the necessary collision with oxygen and/or substrate. Those compounds that can form long-lived ($\sim 10^{-6}$

seconds) excited states (e.g., triplet states) by intersystem crossing from their excited singlet states are effective as photodynamic agents.

## 9-3. MECHANISMS OF PHOTODYNAMIC ACTION

An understanding of the mechanism of photodynamic action first requires a knowledge of the processes of excitation and deexcitation of the dye. The probable photochemical events that take place when a suitable dye ($D$) is exposed to visible light are:

$D + hv \rightarrow {}^1D$      Absorption of visible light to give dye molecules in the first excited singlet state ($^1D$).

${}^1D \longrightarrow D$      Internal conversion and radiationless deexcitation to ground state.

${}^1D \longrightarrow D + hv_f$      Fluorescence.

${}^1D \longrightarrow {}^3D$      Intersystem crossing to long-lived triplet state ($^3D$).

${}^3D \longrightarrow D + hv_p$      Phosphorescence.

Photodynamic action appears to proceed via the triplet state ($^3D$) of the dye.

Several mechanisms by which an excited dye molecule ($^3D$) may cause the oxidation of a substrate molecule ($S$) have been demonstrated. Five such mechanisms are as follows:

(1) $D + hv \longrightarrow {}^1D \rightarrow {}^3D$    (3) $D + S \longrightarrow D \cdot S$
$\quad\ {}^3D + O_2 \rightarrow DO_2$             $D \cdot S + hv \longrightarrow {}^1D \cdot S \rightarrow {}^3D \cdot S$
$\quad\ DO_2 + S \rightarrow D + SO_2$         ${}^3D \cdot S + O_2 \longrightarrow D + SO_2$

(2) $D + hv \longrightarrow {}^1D \rightarrow {}^3D$    (4) $D + S + O_2 \longrightarrow D \cdot S \cdot O_2$
$\quad\ {}^3D + S \longrightarrow D + {}^3S$          $D \cdot S \cdot O_2 + hv \rightarrow D + SO_2$
$\quad\ {}^3S + O_2 \longrightarrow SO_2$

(5) $D + hv \longrightarrow {}^1D \rightarrow {}^3D$
$\quad\ {}^3D + {}^3O_2 \longrightarrow D + {}^1O_2$
$\quad\ {}^1O_2 + S \longrightarrow SO_2$

In the first case, the dye is activated and oxidized to a "photoperoxide" which in turn oxidizes the substrate. In the second case, the excited dye molecule transfers excitation energy to the substrate which then is directly oxidized by molecular oxygen. In the third case, the dye first combines with the substrate and then upon illumination the dye-substrate complex becomes excited and sub-sequently reacts with oxygen to yield free dye and oxidized sub-strate. In case four, the dye, substrate and oxygen form a complex which, upon illumination, yields oxidized substrate. In case five, the dye in its excited triplet state reacts with triplet oxygen (the normal ground state of oxygen) to form singlet oxygen which is the species that oxidizes the substrate.

The mechanism that has enjoyed the greatest favor until recent years is mechanism one, originally put forward by Schenck. More recently, however, mechanism five, originally put forth by Kautsky in 1931 has achieved favor and much work on singlet oxygen is in progress throughout the world. It should not be overlooked, how-ever, that all five mechanisms have basis in experimental fact. One must therefore conclude that no one mechanism fits all cases. It is not unreasonable to assume that different dyes may act by differ-ent mechanisms on a given substrate, or that the same dye might act by different mechanisms upon different substrates. It should be stressed, however, that in all cases regardless of the subsequent events, the primary event of photosensitization is the absorption of light by the photosensitizing molecule. An obligatory subsequent event in photodynamic action is the consumption of molecular oxygen.

Although photodynamic action has been studied for many years, we still have very little understanding of the fundamental, molec-ular mechanisms involved. The basic approach to date has been the empirical testing of many types of dyes in a multitude of systems in order to gain some information as to the chemical requirements for an active dye. The analytical approach to the problem has largely been one of determining the chemical kinetics of the reaction. There is now a long list of photodynamically'active dyes belonging to a wide range of chemically different subclassifications and we have some knowledge of their kinetics. Phase one, or the empirical era, in the evolution of the discipline of photodynamic action would seem to be drawing to a close. What is now required are detailed studies on the molecular mechanisms involved.

9-4.  Photodynamic Action on Cells

Many types of photodynamic phenomena have been studied at the cellular level, for example, delays and abnormalities in cell division, production of mutations, effects on metabolism, killing and so forth. Little is known, however, about the basic chemical mechanisms involved in these photodynamic processes.

Figure 9-3 shows typical results for the killing of *E. coli* B by visible light in the presence of different concentrations of acridine orange. Strains of *E. coli* which show marked differences in sensitivity to UV inactivation have shown only slight differences in response to photodynamic inactivation. Thus, *E. coli* $B_s$ and B/r which differ in sensitivity to UV by about a factor of 100 (Figure 6-7) only show at most a twofold difference in sensitivity to

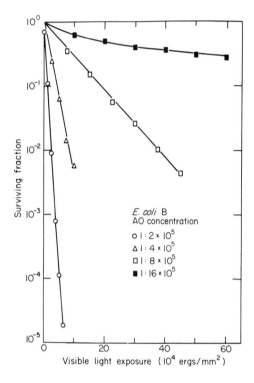

FIGURE 9-3. The survival of *Escherichia coli* B when irradiated with visible light in the presence of several different concentrations of acridine orange (AO). [R. B. Uretz, unpublished data (1964); but see *Radiation Res.* **22**, 245 (1964) (abstr.).]

acridine orange (and only about a three-fold difference in sensitivity to X-rays). Earlier reports showed no differences between the sensitivity of *E. coli* B$_s$ and B/r in the presence of acridine orange, and, in fact, the majority of published reports indicate little or no repair of photodynamic effects. One may wonder, therefore, whether the similar sensitivity of *E. coli* B$_s$ and B/r to photodynamic inactivation is due to the fact that both strains have similar sensitivities to inactivation (i.e., little or no repair) or a similar resistance to inactivation (i.e., a similar ability to repair). How might these two alternatives be distinguished?

Yeast cells that have been irradiated with X-rays under nitrogen are more sensitive to subsequent treatment with visible light and acridine orange than are cells first irradiated with X-rays under oxygen. In fact, there was no interaction of the two treatments (no synergism) at all under aerobic conditions or if the order of the radiations was reversed. It was concluded that X-irradiation under nitrogen produces nonlethal lesions (whose nature is unknown) which render the cells more susceptible to photodynamic inactivation. It is also possible that this phenomenon is due to an increased permeability of the cell to the dye.

Irradiation of thymine dimers in aqueous solution with visible light in the presence of uranyl acetate results in the liberation of thymine and some unidentified products. The photoreactivation (Section 7-3) of UV-irradiated *E. coli* B was about 10% more efficient in the presence of uranyl acetate. The photoreactivation of UV-irradiated bacteria can also be enhanced by the presence of hydrogen peroxide. The use of these compounds to enhance photoreactivation is limited, however, by their toxicity to the cell.

Bacteria and other cells are susceptible to mutation production and to inactivation by visible light *in the absence of added dye* (but at greatly reduced efficiency). Since mutations are produced, it is assumed that they result from damage to DNA. Mutation production could result either from a photodynamic response due to the presence of endogenous sensitizers (such as riboflavin) or to nonphotodynamic triplet energy transfer from ketones to the pyrimidines, as discussed in Section 4-5. Cells containing heavily pigmented cell walls are protected from visible light inactivation, as compared to nonpigmented cells of the same strain.

9-5. PHOTODYNAMIC ACTION ON VIRUSES

Serologically related bacteriophages exhibit similar susceptibility to visible light in the presence of a given dye, but there are marked differences in sensitivity from one serological grouping to another. The T-even phages are generally quite resistant to photodynamic inactivation while the T-odd phages are quite sensitive. Since the T-even phages are more susceptible to osmotic shock than are the T-odd phages, it is supposed that the protein coats of the T-even phages are less permeable to dyes than are the coats of the T-odd phages. These differences in sensitivity of viruses to photodynamic inactivation can be used to destroy a sensitive contaminant in the presence of a resistant virus.

Mature polio virus is not sensitive to light in the presence of photosensitizing dyes. Apparently the protein coat of the mature virus is impermeable to dyes. If, however, the virus is grown in cells in the presence of dye (in the dark) the resulting viruses are photosensitive, due to the binding of dye molecules by the RNA during virus replication and its consequent incorporation into the mature virus. The fact that animal viruses containing RNA are more resistant to photodynamic inactivation than are those containing DNA may possibly be explained on the basis of permeability, although it might also be explained on the basis of a differential susceptibility of the two kinds of nucleic acids to this type of inactivation. There is currently no evidence to distinguish between these two possibilities.

Photodynamically inactivated viruses frequently retain their antigenic properties and thus can be used as vaccines.

We have noted that bacterial cells show a marginal capacity to repair photodynamic damage inflicted upon themselves (Section 9-4). They appear to be unable to repair photodynamically damaged (acridine orange) phage T7. This is evidenced by the fact that there was no difference in viability whether the damaged T7 phage were plated on E. coli B or $B_s$, although strain B (but not $B_s$) can repair UV-damaged phage (host-cell reactivation, Section 7-4).

9-6. PHOTODYNAMIC ACTION ON PROTEINS AND NUCLEIC ACIDS

Several approaches have been used to study photodynamic effects on nucleic acids, proteins and their derivatives. The loss of biological activity of transforming DNA or of enzymes has been measured under a variety of conditions. Attempts have been made to determine the changes in the physical state of these polymers that cause biological inactivation. Studies have also been concerned with determining which specific base or amino acid is destroyed as a consequence of photodynamic inactivation.

Studies using methylene blue (Figure 9-2) have shown that tyrosine, tryptophan, histidine, methionine and cystine are particularly susceptible to photodynamic alteration, with histidine being the most sensitive. The chemical changes produced in these amino acids are largely unknown. It was observed that the rate of inactivation (in the presence of methylene blue) of the enzyme phosphoglucomutase was correlated in part with the rate of destruction of the histidine groups on the surface of the enzyme. Peptide bonds do not appear to be ruptured by photodynamic action.

Studies on DNA monomers indicate that deoxyguanylic acid is very sensitive to alteration in the presence of methylene blue with thymidylic acid only slightly sensitive and deoxyadenylic acid and deoxycytidylic acid not affected at all under the conditions used. This selectivity is also carried over to DNA. Most of the photodynamic dyes tested show a preferential attack on guanine. The products arising from these base alterations, however, have not been chemically identified.

Acridine orange is a potent lethal agent for cells in the presence of visible light (Figure 9-3). Its action is apparently directed towards DNA but does not show any preferential attack on the bases. Recent studies on tobacco mosaic virus RNA, however, do indicate some preferential attack on the guanine residues by acridine orange plus visible light but at greatly reduced efficiency as compared to methylene blue. The major chemical effect of acridine orange thus far observed on isolated DNA is that of chain breakage. The cross-linking of DNA and protein in the presence of this dye (and also methylene blue) has been observed both *in vivo* and *in vitro*. A good correlation has been observed between the enhanced killing of *E. coli* in the presence of methylene blue (relative to acridine orange) and the enhanced cross-linking of DNA and protein in the presence of methylene blue (relative to acridine orange) (Figure 9–4).

When peptides are dissolved in 98–100% formic acid or acetic acid in the presence of proflavin (Figure 9-2) only methionine and tryptophan are photooxidized. By treatment of the oxidized peptide

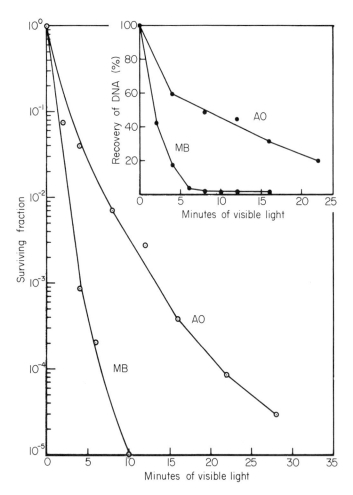

FIGURE 9-4.   The killing of *E. coli* B/r,T⁻ and the cross-linking of DNA and protein by visible light in the presence of acridine orange (3.4 $\mu$g/ml) or methylene blue (4 $\mu$g/ml). Stationary phase cells were washed and suspended in 0.1M phosphate buffer (pH 6.8) and dye was added to the final concentrations indicated. Solutions were irradiated with a 150 W General Electric Floodlight (PAR-38) through 1 inch of water in a heavy pyrex dish. Viability was determined on nutrient agar plates. The DNA was isolated after lysing the cells with sodium lauryl sulfate. [K. C. Smith, unpublished observations (1966).]

with mercaptoethanol the methionine can be reduced leaving only the tryptophan residues altered. This procedure would appear to be applicable to the study of the importance of methionine and/or tryptophan residues in certain biologically active peptides.

In the study on the photodynamic effect of riboflavin on the amino acid acceptor activity of *E. coli* transfer RNA, phenylalanine and lysine acceptor activity were lost in a first-order process with the phenylalanine acceptor activity being the most sensitive.

## 9-7. PHOTOSENSITIZED REACTIONS NOT REQUIRING OXYGEN

Up to this point, we have only discussed photosensitized reactions that require molecular oxygen (photodynamic action) but it should be pointed out that dye-sensitized photochemical oxidations involving biological systems can occur in the absence of molecular oxygen provided some other appropriate electron and/or hydrogen donor is present. Thus (using the symbols defined in Section 9-3):

$$D + h\nu \rightarrow {}^1D \rightarrow {}^3D$$
$$^3D + S + H_2O \rightarrow DH_2 + SO$$

Riboflavin illuminated under anaerobic conditions is reported to undergo such a reduction, with water serving as the ultimate hydrogen donor.

The furocoumarins (psoralens) such as 8-methoxypsoralen (Figure 9-5) sensitize bacteria to killing by visible light under anaerobic conditions, and in fact, the presence of oxygen *decreases* the efficiency of the dye-sensitized reaction. Another peculiar response of this dye is that in the temperature range of 0°C–39°C, there is less killing as the temperature is increased, whereas photodynamic dyes show an increase in killing as the temperature increases in this range. The furocoumarins have recently been shown to react with

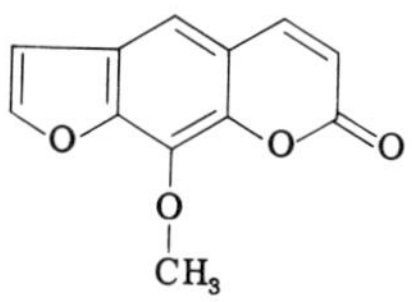

FIGURE 9-5. 8-Methoxypsoralen.

the 5–6 double bond of thymine and cytosine to yield photoproducts containing a cyclobutane ring between the two reactants. *E. coli* B/r is much more resistant to ultraviolet radiation (and X-rays) than *E. coli* B and similarly it is more resistant to killing by 8-methoxypsoralen. The similar sensitivities of *E. coli* B and B/r to killing by the photodynamic action of acridine orange suggests that the chemical changes produced by photodynamic action are quite different from those produced by the furocoumarins (psoralens).

9-8.   PHOTOSENSITIZED REACTIONS IN WHOLE ANIMALS

In some geographical areas the syndrome produced in mammals by photodynamic action has been a real problem for farm animals until weeds containing photosensitizing compounds were removed from the area. Photosensitization leading to erythema and edema and more severely to necrosis and sloughing of the exposed areas (and even death) has been observed. The generalized toxic reactions probably result when compounds in the skin are altered by photo-sensitized reactions and are then carried throughout the body by the circulatory system. Photosensitizing compounds can be introduced into animals by three mechanisms: (1) the ingestion of a sensitizer in the diet, (2) the production of a sensitizer by the animal due to aberrant pigment (e.g., porphyrin) synthesis, or (3) the diversion of a compound that is normally absorbed and excreted (e.g., chlorophyll) to the peripheral circulation because of a liver dysfunction.

In man, the same mechanisms apply but photosensitization most frequently arises as a consequence of medication with fluorescent drugs. Photosensitization can also arise as a result of contact with chemicals if conditions are favorable for their penetration through the skin (i.e., additives in soaps and lotions).

GENERAL REFERENCES

N. T. Clare, Photodynamic action and its pathological effects. *In* "Radiation Biology" (A. Hollaender, ed.), Vol. 3, p. 693. McGraw-Hill, New York, 1956.

L. Santamaria and G. Prino, The photodynamic substances and their mechanism of action. *Res. Prog. Org.-Biol. Med. Chem.* 1, 259 (1964).

A. D. McLaren and D. Shugar, "Photochemistry of Proteins and Nucleic Acids," pp. 156 and 313. Pergamon Press, Oxford, 1964.

M. I. Simon, Photosensitization. *In* "Photobiology, Ionizing Radiation" (M. Florkin and E. H. Stotz, eds.), p. 137. Elsevier, Amsterdam, 1967.

J. D. Spikes, Photodynamic action. *In* "Photophysiology" (A. C. Giese, ed.), Vol. 3, p. 33. Academic Press, New York, 1968.

J. D. Spikes and R. Livingston, The molecular biology of photodynamic action. *Advan. Radiation Biol.* 3, (1969) (in press).

# 10

# Comparison of UV and Ionizing Radiation

## 10-1.  INTRODUCTION

Our knowledge of the molecular mechanisms leading to cell death after X-irradiation lags far behind that for UV irradiation. One explanation is that more effort has been expended in recent years on the problems involved with UV irradiation. Another explanation is that molecular excitations produced by UV are more specific and are more easily studied than random ionizations produced by X-rays within molecules. Ionizing radiation is also less selective than UV as to types of molecules that it attacks. In spite of this, not all compounds within a cell have the same radiological importance.

Many characteristic features of the response of living cells to ionizing radiations are expressions of the ways in which the cells are organized. If the same small proportion of each class of molecules is destroyed indiscriminately by ionizing radiation, the loss of a few molecules from those classes which occur in high multiplicity would not be noticed, and the observed response would be traceable to those which are unique. It is mainly this line of reasoning that leads one to the conclusion that DNA must be the primary target for radiation effects.

## 10-2. THE TARGET OF PRIMARY RADIOBIOLOGICAL IMPORTANCE IS DNA

If a selective aberration can be introduced into one class of molecules within a cell and if such cells are then found to be much more radiation sensitive it is reasonable to suspect that this class of macromolecules is the primary target for the lethal effects of radiation. If cells are grown in the presence of pyrimidine (or purine)

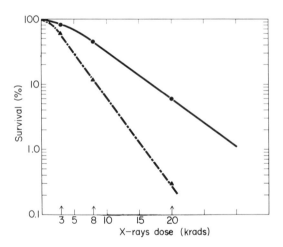

FIGURE 10-1.   The effect of X-irradiation on the survival of *E. coli* B,T⁻, strain W4516, that had been grown either on thymine or bromouracil. ●—●, Thymine; ▲ --- ▲, 5-bromouracil (7 hour cultures). [From H. S. Kaplan, K. C. Smith, and P. A. Tomlin, *Radiation Res.* **16**, 98 (1962).]

analogs under conditions permitting their assimilation but not in-corporation into DNA, such cells show normal radiation sensitivity. However, if base analogs (e.g., 5BU) are allowed to replace the normal bases in the DNA of cells, then these cells show a greater sensitivity to killing by X-rays, suggesting that the DNA is the primary target of ionizing radiation leading to reproductive death (Figure 10-1; see also Figure 4-16 for response to UV).

More direct evidence that the initial radiation damage occurs within the DNA molecule was provided by studies with bacterial transforming DNA, which can be irradiated either *in vivo* (in the bacterial cell) or *in vitro* (after extraction). Transforming DNA, into which 5BU was incorporated, was found to be more sensitive than normal DNA to X-rays (and to ultraviolet light), whether irradiated *in vivo* or *in vitro*. It may thus be concluded with some assurance that the primary radiobiochemical lesions leading to cell death occur in the DNA.

10-3. THE EFFECTS OF DNA BASE COMPOSITION, DNA
CONTENT, PLOIDY, AND STRANDEDNESS OF THE DNA
ON THE RADIATION SENSITIVITY OF CELLS

The use of analogs is not necessary in order to obtain correlative evidence on the radiobiological importance of DNA. The natural bases are themselves analogs of each other. Thus, cytosine is really an analog of thymine and adenine is an analog of guanine. There are many species of bacteria whose DNA differs markedly in the rela-tive content of adenine and thymine relative to guanine and cyto-sine (e.g. *Pseudomonas aeruginosa,* 67% GC; *Escherichia coli,* 50% GC; *Bacillus cereus,* 35% GC). If the target for radiation is the DNA and if the radiation sensitivity of the DNA is modified by analog substitution, then it is quite possible that bacteria with DNA of different base compositions might also differ in some systematic way in their sensitivity to the lethal effects of radiation.

Experiments performed with X-rays showed that the sensitivity to killing of several strains of bacteria could be correlated with the guanine and cytosine content of their DNA (Figure 10-2). There was a linear increase in the sensitivity of different bacteria with increasing contents of guanine and cytosine. The data in the litera-ture for the radiation susceptibility of the purines and pyrimidines are not in agreement (see Section 10-7) so that we currently have

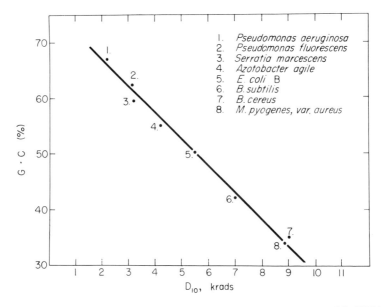

FIGURE 10-2.　The relation between the guanine-cytosine (GC) content of the DNA of different bacteria and the $D_{10}$ in krads, where $D_{10}$ is the dose which reduces the surviving fraction to 10% on the exponential part of the survival curve. [From H. S. Kaplan and R. Zavarine, *Biochem. Biophys. Res. Commun.* 8, 432 (1962).]

no clear chemical explanation for the apparent greater susceptibility of guanine and cytosine to X-rays *in vivo*.

When similar experiments were performed using ultraviolet light (see Section 4-10), it was found that the sensitivity of killing could be correlated in a linear manner with the adenine and thymine content of the bacterial DNA. The bacteria with highest content of adenine and thymine were the most sensitive. This relationship would seem to be adequately explained by our present knowledge of the importance of thymine photoproducts in the UV inactivation of DNA.

The total nucleotide content of several plant, animal and bacterial viruses, haploid and diploid microorganisms and of diploid mam-

doses (Figure 10-3). The values for diploid cells lie on one iso-sensitivity line (top line) indicating that as the DNA content increases so does the sensitivity to X-rays. The values for double-stranded DNA bacteriophages and haploid microorganisms lie on a second isosensitivity line (two middle lines) shifted about tenfold in the direction of increased sensitivity from that for the diploid cells. Data for RNA-containing animal and plant viruses and two viruses containing single-stranded DNA lie on a third isosensitivity line (bottom line) about tenfold more sensitive than the haploid DNA line. These data suggest that radiosensitivity is in part dependent upon ploidy, nucleic acid content, and strandedness.

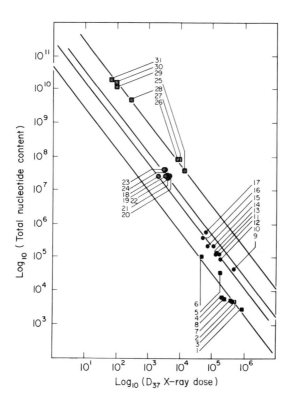

FIGURE 10-3.  Forty-five-degree lines fitted to data points representing various cells and viruses of four classes: ■, RNA and single-stranded DNA viruses; ●, double-stranded DNA viruses; , haploid bacteria and yeast; , mammalian and avian cells and diploid yeast. The numbers indicating the different cell lines are identified in the original publication. [From H. S. Kaplan and L. E. Moses, *Science* **145**, 21 (1964).]

10-4. THE BIOCHEMICAL EFFECTS OF X-IRRADIATION

If DNA is the target, then how are the effects of its alteration manifested? The older literature is filled with papers on the inhibition of DNA synthesis in the organs of animals following whole-body X-irradiation. In these experiments the extent of incorporation of injected radioactive precursors into the DNA of several tissues following X-irradiation was compared with that for unirradiated animals. Although the irradiated animals showed a reduction in incorporation of DNA precursors, most of these data have now been shown to be artifactual. The apparent alterations in DNA synthesis were largely due to precursor pool dilutions by metabolites from dead cells, and to mitotic inhibition rather than to direct effects upon DNA synthesis. This sums up an immense amount of research prior to 1958; data which were difficult to interpret and which led to extensive arguments in the literature. This problem has also been investigated using the simpler system of tissue culture cells. In 1961, Till reported only a 15% decrease in the rate of DNA synthesis after 2000 Roentgens (R) a dose sufficient to kill 99.9% of the cells. He concluded that the effects of X-irradiation on DNA synthesis in mammalian cells in tissue culture are indirect consequences of the inhibition of mitosis.

The main action of X-irradiation on DNA in mammalian cells, therefore, does not appear to be the direct inhibition of DNA synthesis. However, one must appreciate the important differences between bacterial and mammalian genomes. The bacterial chromosome, with which most of our discussion has been concerned, consists of just one molecule of double-stranded DNA. In contrast, a mammalian chromosome contains several molecules of DNA and a total content of DNA about 100 times that of the bacterial chromosome. Furthermore, mammalian cells contain many different chromosomes. DNA replication occurs simultaneously at many sites along the mammalian chromosome while it usually occurs at only one site on the bacterial chromosome. Inhibition of replication in a bacterial chromosome would be reflected as a total inhibition of DNA synthesis. Although the inhibition of replication at one of the sites on a mammalian chromosome would probably be lethal, its effect upon overall DNA synthesis would not even be detected by the usual analytical procedures. The relative importance of mitotic inhibition on DNA synthesis compared to the direct effects of X-irradiation on synthesis remains an intriguing but unresolved problem.

*In vitro* studies have shown that the enzymes involved in DNA synthesis are quite radiation resistant. The ability of DNA to act as a primer for the *in vitro* synthesis of both DNA and RNA has also been investigated. Using quite impure preparations of DNA polymerase, earlier investigators found little or no effect of X-irradiation on the ability of DNA to act as a primer for the *in vitro* synthesis of DNA. However, using more purified enzyme preparations, a significant effect of X-irradiation upon the priming ability of DNA for both DNA and RNA synthesis has been demonstrated *in vitro*. These results may possibly be explained in terms of repair enzymes present in the crude enzyme extracts.

The inhibition of the function of DNA by radiation has been demonstrated in a cell-free system that synthesizes the enzyme $\beta$-galactosidase. This system requires DNA, messenger RNA, ribosomes, transfer RNA, amino acids and several enzymes. If this protein synthesizing system is isolated from irradiated cells, it will not make $\beta$-galactosidase, but if purified DNA from unirradiated cells is added, the system will synthesize the enzyme. From our current knowledge of protein synthesis we know that DNA codes for the formation of messenger RNA and that this RNA carries the necessary information to the ribosomes for the formation of a specific protein. If a faulty messenger RNA or no messenger RNA is made because the DNA is damaged, then the system cannot make the enzyme.

Another mechanism by which irradiation might interfere with the function of DNA derives from an investigation of the effect of X-irradiation on the ability of deoxyribonucleoprotein to act as a primer for RNA synthesis *in vitro*. When pure DNA is irradiated, its activity as a primer for RNA synthesis steadily declines with dose. When deoxyribonucleoprotein is irradiated, there is an initial decrease in priming activity followed by a sharp increase in activity. The interpretation of this response was that at the lower doses of X-irradiation the typical inactivation curve for free DNA was obtained (the sample of DNA was not completely covered with protein). At the higher doses the protein became dissociated from the DNA and thereby exposed more DNA that could be used to code for RNA synthesis. At still higher doses the exposed DNA should become inactivated.

It has been suggested that the proteins found associated with the DNA may act as repressors and cover up regions of the DNA that are not to be transcribed into messenger RNA in particular differentiated cells or at a particular time in the generation cycle of a given cell. If irradiation caused the release of some of this protective protein coat, the exposed sections of the DNA could then be used to prime the synthesis of messenger RNA. The enzymes that would be subsequently synthesized would be inappropriate and possibly detrimental to the normal function of the cell. It should be stressed that there is at present no direct evidence to support this hypothesis.

In discussing the repair of UV damage in bacteria (Section 7-4) we have commented that DNA is broken down after UV irradiation in certain radiation-resistant strains (e.g., *E. coli* B/r) but not in certain radiation-sensitive strains (e.g., *E. coli* $B_{s-1}$). This is explained by the presence of the excision-repair mechanism for the repair of UV-induced damage in *E. coli* B/r and its absence in *E. coli* $B_{s-1}$. A characteristic response of bacterial cells to X-irradiation, however, is a much enhanced breakdown of DNA (compared to UV) for the same killing dose. For an X-ray dose yielding the same survival there is about 35% breakdown of the DNA in *E. coli* B/r and an 80% breakdown in *E. coli* $B_{s-1}$. Clearly the bulk of this breakdown cannot be attributed to repair processes; in fact, the ability to terminate this breakdown would seem to be a characteristic of the more radiation resistant strain. What precisely triggers the DNA breakdown, what stops it, and how the damage is repaired remain intriguing but unanswered questions.

10-5.  GENERAL DISCUSSION OF RADIATION CHEMISTRY

*5A. Introduction*

Although both X-rays and ultraviolet rays are composed of photons, the energy of the X-ray photon is about $10^4$ greater than the energy of the photon of ultraviolet light. The major consequence of the absorption of a photon of X-rays is not molecular excitation as it is for UV (although excitations are produced), but is rather the production of an ionization as the consequence of the loss of an orbital electron from the target atom or molecule. The ejected electron then proceeds to cause a series of ionizations by electrostatic repulsion in other molecules along its trajectory until all of its energy is dissipated or until it is recaptured to produce an excited molecule.

On the average, one ionization occurs in a system for each 32.5 ev absorbed. However, the actual amount of energy needed to eject an electron from a molecule (ionization potential) ranges only from 10–25 ev. The extra energy absorbed can result in excited molecules. There is a high probability that ionized molecules will undergo permanent chemical alteration. We know from our previous discussion of quantum chemistry (Chapter 3) that the energy of excited molecules can frequently be dissipated in a harmless manner. It is for this reason that one often finds statements to the effect that excitations are not biologically important. This is an unwarranted oversimplification. Ionizations are easy to measure while molecular excitations are more difficult. Hammond and co-workers have measured the production, by gamma radiation, of singlet and triplet states in chemical systems previously well characterized in photochemical studies. These studies stress that molecular excitations are important in high-energy radiation chemistry.

The unit of X- or $\gamma$-radiation is the Roentgen (R) and is based on the ionization that these radiations produce in air. An *exposure dose* of one R results in $2.584 \times 10^{-4}$ coulomb/kg of dry air, or 1 esu/cm$^3$ of air at standard temperature and pressure. The *rad* is a unit of *absorbed dose* for ionizing radiation. One rad is 100 ergs absorbed per gram of any substance. In water or soft tissue and for radiation whose energy is between 100 kev and 3 Mev the absorbed dose per roentgen is about 0.95 rad.

In photochemistry one determines the sensitivity of a compound to undergo photochemical alteration by measuring its quantum yield (Section 1-2). The analogous parameter in radiation chemistry is the G value, i.e., the number of molecules affected per 100 electron volts absorbed (see Section 10-7).

## 5B.  Direct and Indirect Effects

There are two distinct mechanisms by which a chemical change can be brought about by ionizing radiations. One is direct action. Here the target molecule becomes ionized or excited by the passage through it of a photon, an electron or other atomic particle. The second case is called indirect action and the molecule under investigation does not absorb the energy of the radiation directly but rather this energy is transferred to it from another energized molecule.

Since the most prevalent molecule in an aqueous system is water, it is not surprising that the process of indirect action plays a significant role in the inactivation of solute molecules. The ionization of a water molecule leads to the formation of free radicals (see Section 10-6) which readily attack the solute molecules.

The chief method for distinguishing between direct and indirect action is by dilution of the solute. If the action of the radiation on the solute is indirect the number of molecules of solute inactivated by a given dose of radiation will be constant since a fixed number of free radicals is produced in water by a given dose of radiation. The number of solute molecules inactivated will decrease with increasing concentrations of solute. If the action is direct, the number of molecules inactivated by a given dose of radiation will vary with the concentration and the percentage inactivation will be constant for a given dose (Figure 10-4).

When an enzyme or transforming DNA is irradiated in solution, indirect effects mediated through the solvent play the principal role in the inactivation process. When these compounds are irradiated dry, direct effects play the major role. There is as yet no clear understanding of the relative importance of direct and indirect effects *in vivo*.

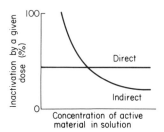 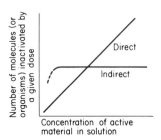

FIGURE 10-4. The effect of dilution of solute upon the direct and indirect effects of X-irradiation on the solute. The results can be expressed either as the percentage inactivation of the whole solution or as the number of macromolecules inactivated. (From Z. M. Bacq and P. Alexander, "Fundamentals of Radiobiology." Pergamon Press, Oxford, 1961.)

*5C. Chemical Protection Against Radiation Damage*

The free radicals produced in water are highly reactive, and interact readily with a wide variety of different molecules. If a second substance is added to a solution it will compete for these radicals with the target molecule (the enzyme or other biologically active molecule) and thereby reduce the extent of the inactivation of the target molecule. Radical scavenging thus forms one basis for chemical protection against the indirect effects of radiation. Certain sulfur-containing compounds have been found to be particularly effective as protective agents (e.g., cysteine, cysteamine, etc.).

Reactions leading to protection against the direct action of radiation have also been reported. If the first chemical change that occurs in a target molecule is reversible there may be a short time during which the molecule can react with a protector substance in such a way that the target molecule is regenerated. An example of such repair could involve a transfer of hydrogen (H) from the SH group of the protector molecule (R'SH) to the damaged target molecule (R·).

$$RH \xrightarrow{\ h\nu\ } R\cdot + H\cdot \qquad \text{(Destruction of target molecule)}$$
$$R\cdot + R'SH \rightarrow RH + R'S\cdot \qquad \text{(Repair by protector molecule)}$$

Radiation protection can also be built into a molecule. The presence of an aromatic ring in an aliphatic chain greatly increases the radiation resistance of that chain. Thus, the energy to produce a break in a chain of polyisobutylene is 18 ev, but this is increased to 35 ev in a copolymer of which 20% is composed of styrene rings, and to 100 ev if the styrene content is raised to 80%. This is presumed to be due to the migration of the energy from the aliphatic chain to the aromatic rings and the subsequent nondestructive dissipation of the energy as heat or light.

*5D. Oxygen Effect*

When biological systems are irradiated with ultraviolet light, the response is independent of whether the system is irradiated in the presence or absence of oxygen. This is not true for X-rays. There is an increase in sensitivity of the order of threefold in the presence of oxygen as compared to nitrogen. Both the types and relative proportions of free radicals produced in oxygenated water differ significantly from those produced in the absence of oxygen (Section 10-6).

Oxygen also enhances the direct action of radiation. Molecular oxygen can add to a radical produced in a target molecule to give a product (i.e., peroxide) that cannot undergo spontaneous chemical repair (Section 10-6).

### 5E.  Types of Damage Expected

Almost all types of organic chemical reactions can be produced by ionizing radiation: oxidations, reductions, additions, deletions, chain breaks, and internal rearrangements.

Chemical effects much less severe than these, however, may be adequate to inactivate large molecules such as proteins. The structure of a protein must be described at several different levels. The primary structure refers to the order in which the different amino acids are strung together to make up the polypeptide chain. The secondary structure is the configuration adopted by the chain, that is, whether it is fully extended or whether it is coiled in some way. The tertiary structure is concerned with the way in which the chain is folded upon itself. Since the maintenance of the three dimensional structure of an enzyme is required for its enzymatic activity, the simple breakage of hydrogen bonds by radiation may be sufficient to disorganize the tertiary structure and therefore inactivate the enzyme. Under appropriate environmental conditions some enzymes may reform their normal tertiary structure and thereby regain activity.

### 5F.  Energy Migration

Adjacent atoms in a molecule are covalently bonded due to the sharing of electrons. If an atom in a small molecule becomes ionized, due to the loss of an electron that binds it to its neighbor, it would surely cause disruption of the molecule. The tertiary structure of a very large molecule such as a protein, however, would be expected to hold the atom in place, at least for a fraction of a second. In response to the positive charge on the carbon atom (due to the loss of the electron), there would be a strong tendency for an electron from a neighboring atom to move over and neutralize the first atom. The atom that had just donated the electron would then assume the positive charge. In this manner, the positive charge could traverse the entire protein molecule. If the polypeptide chain could normally assume an alternative configuration, but is held in place by one or more weak bonds, the opportunity to change this configuration and perhaps inactivate the molecule might arise by bond breakage resulting from charge migration.

The process of energy transfer can thus make a large biologically active molecule hypersensitive to radiation because a hit anywhere in the molecule may cause the inactivation of the small sensitive area in the molecule. Alternatively, however, we have seen the effect of introducing styrene molecules into an aliphatic chain. In this case, energy transfer to the aromatic rings makes the whole molecule much more resistant to destruction by ionizing radiation (Section 10-5,C).

10-6. RADIATION CHEMISTRY OF WATER

Most of the radiation effects on solutes in aqueous systems are indirect effects caused by free radicals produced in water by ionizing radiation. The response of water to ionizing radiation is different in the presence and in the absence of oxygen.

In the absence of oxygen the following reactions occur:

$$
\begin{aligned}
H_2O + h\nu &\rightarrow H_2O^+ + e^- \\
H_2O^+ + H_2O &\rightarrow H^+ + H_2O + OH\cdot \\
e^- + H_2O &\rightarrow H_2O^- \\
H_2O^- + H_2O &\rightarrow OH^- + H_2O + H\cdot \\
\hline
H_2O + h\nu &\rightarrow H\cdot + OH\cdot
\end{aligned}
$$

Thus, optimally one reducing radical ($H\cdot$) and one oxidizing radical ($OH\cdot$) will be generated in the absence of oxygen. One might naturally expect that the hydrogen radical ($H\cdot$) would undo all of the oxidations produced by the hydroxyl radicals ($OH\cdot$). The formation of the $OH\cdot$ radical is mandatory as the primary event in the ionization process, but the ejected electron does not necessarily result in the production of a hydrogen radical ($H\cdot$). The electron can simply be recaptured (producing an excited molecule) or it can enter into other reactions leading to the formation of products with decreased reducing efficiencies (e.g., $H_2^+$). If two hydrogen radicals ($H\cdot$) interact, they form molecular hydrogen ($H_2$) which is chemically inert. Whereas if two hydroxyl radicals (OH·) interact they form hydrogen peroxide ($H_2O_2$) which is still an active oxidizing agent. In fact, a notable feature of radiation chemistry in aqueous solutions is that oxidizing reactions predominate, even in the absence of oxygen.

At neutral pH the reducing species produced by the radiolysis of water is not the hydrogen radical but rather the hydrated electron ($e_{aq}^-$). Evidence for the presence of hydrated electrons is based upon the observation that the radiation-induced reducing species have a negative charge while hydrogen radicals have no charge. The reducing species produced by the X-irradiation of water also reacts somewhat differently than does H·. Hydrogen radicals can be produced by microwave discharge in hydrogen gas and these can then be bubbled through a test solution. All reducing reactions at neutral pH, however, are not necessarily due to $e_{aq}^-$. Hydrated electrons can react with certain chemical species to liberate H· and these can then attack the solute molecules.

$$e_{aq}^- + H_2PO_4^- \rightarrow H· + HPO_4^{2-}$$

This indicates that buffers cannot be considered inert in a system to be irradiated. Formaldehyde is rapidly attacked by H· but not by $e_{aq}^-$. If the addition of formaldehyde does not alter the yield of reduced products during irradiation it would suggest that the reducing species present was $e_{aq}^-$.

In the presence of oxygen the following reactions occur:

$$
\begin{array}{ll}
H_2O + h\nu & \rightarrow H_2O^+ + e^- \\
H_2O^+ + H_2O & \rightarrow H^+ + H_2O + OH· \\
e^- + O_2 & \rightarrow O_2^- \\
O_2^- + H_2O & \rightarrow OH^- + HO_2 \\
\hline
H_2O + O_2 + h\nu \rightarrow HO_2 + OH·
\end{array}
$$

Thus, in oxygenated systems, the peroxy radical ($HO_2$) is formed instead of the H· radical (or $e_{aq}^-$). The $HO_2$ radical has three times the oxidizing capacity of an OH· radical.

$$
\begin{array}{ll}
RH + HO_2 & \rightarrow R· + H_2O_2 \\
RH + H_2O_2 & \rightarrow R· + OH· + H_2O \\
RH + OH· & \rightarrow R· + H_2O \\
\hline
3RH + HO_2 \rightarrow 3R· + 2H_2O
\end{array}
$$

This explains in part the fact that cells are about three times more sensitive to X-rays in the presence of oxygen than in the absence of oxygen.

An additional consideration is that in the absence of oxygen restoration can occur by the reassociation of an ionized molecule,

$$RH \xrightarrow{\;hv\;} R\cdot + H\cdot$$
$$R\cdot + H\cdot \longrightarrow RH$$

but in the presence of oxygen restoration is blocked.

$$R\cdot + O_2 \rightarrow RO_2^{\cdot} \text{ (peroxy radical)}$$

## 10-7.  RADIATION CHEMISTRY OF THE PURINES AND PYRIMIDINES

As with UV, the purines are more resistant to degradation by ionizing radiation than are the pyrimidines; however, there is no good agreement in the literature on the G values (defined in Section 10-5,A) for the several purines and pyrimidines. A number of factors contribute to this lack of agreement between laboratories and even to the lack of reproducibility within the same laboratory. G values appear to be considered by some to be true physical constants with no qualifying conditions. Nothing could be further from the truth. The sensitivity of a given compound to the effects of X-irradiation depends upon almost every physical and chemical variable that one can think of: concentration, presence or absence of oxygen, presence of other compounds (impurities; buffer, etc.), and the pH of the solution, just to mention a few. Unless each of these parameters is rigorously controlled, there is little point in comparing G values for different compounds.

Some representative G values for the radiation destruction of the deoxyribonucleotides are given in Table 10-1. Since the sum of the G values for guanine and cytosine is not always larger than the sum of the G values for thymine and adenine, it is not possible to use these data to explain the observation (cited in Section 10-3) that bacteria with a high guanine and cytosine content are more easily killed by X-rays than are those with a low guanine and cytosine content.

TABLE 10-1.  G VALUES FOR NUCLEIC ACID BASES

|  | Separate Deoxy-ribonucleotides[a] | Equimolar mixture of Deoxyribonucleotides[b] | DNA[c] | Separate, dry deoxy-ribonucleotides[d] |
|---|---|---|---|---|
| Thymine | 1.65 | 0.47 | 0.33 | 2 |
| Adenine | 0.76 | 0.24 | 0.25 | 2 |
| Sum | 2.41 | 0.71 | 0.58 | 4 |
| Cytosine | 2.00 | 0.34 | 0.23 | 5 |
| Guanine | 0.78 | 0.20 | 0.18 | 3 |
| Sum | 2.78 | 0.54 | 0.41 | 8 |

[a] 0.2 mM at pH 7, 200 kv X-rays; 1 atm oxygen
[b] 50 $\mu$M each at pH 7; 200 kv X-rays; oxygenated
[c] 0.005%; 200 kv X-rays; 1 atm oxygen; G values are for base destruction. From J. J. Weiss, *Progr. Nucleic Acid Res. Mol. Biol.* 3, 103 (1964).
[d] Dry powder under vacuum; Co[60] $\gamma$ source; 15 kR/min; G values are for radical production at 300° K. From A. Müller, *Intern. J. Radiation Biol.* 8, 131 (1964).

If guanosine is exposed to ionizing radiation in the absence of oxygen, a new UV-absorbing material is produced whose structure is 2,4 diamino-5-formamido-6-hydroxypyrimidine (Figure 10-5). This is believed to result from radical attack at the C-8 to N-9 bond in the purine ring. This formamido derivative has been found both free in solution and still attached to ribose. A similar formamido derivative is also observed when adenosine is irradiated at neutral pH in the absence of oxygen. The sugar is attacked by the radiation to a lesser degree than is the base but its alteration results ultimately in the release of free base (adenine and the formamido derivative). In the presence of oxygen the amount of adenosine destroyed increases 2–3 times and is due almost entirely to increased base attack. No UV-absorbing products are observed. Peroxide derivatives of the purines do not appear to be formed as they are for the pyrimidines (see below), or else they

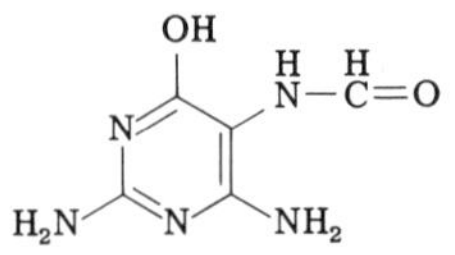

FIGURE 10-5.   2,4-Diamino-5-formamido-6-hydroxypyrimidine.

are so unstable that they have not been detected. Other chemical mechanisms must then result in the saturation and/or cleavage of the purine rings leading to the production of radiation products having no UV absorption.

The pyrimidines form hydroxy-hydroperoxides (and other products) under the influence of ionizing radiation. The radiation products result chiefly from the attack of the 5–6 double bond by H·, OH·, and $HO_2^-$ radicals. The hydroxy-hydroperoxide of uracil (Figure 10-6) is not very stable. It quickly decomposes to form the glycol which then loses a molecule of water to form isobarbituric acid (5-hydroxyuracil). Dihydrouracil, alloxan, alloxantin, dialuric acid and 6-hydroxy, 5-hydrouracil (water addition product) are also formed when uracil is irradiated in solution (Figure 10-6). The hydroxy-hydroperoxide of thymine is quite stable but will slowly decompose to form the glycol derivative. The hydroxy-hydroperoxide of cytosine is extremely unstable and has not yet been isolated. It rapidly changes to the glycol and then can either form 5-hydroxy-uracil by losing both a molecule of water and its amino group or 5-hydroxycytosine if it only loses a molecule of water.

When solutions of uracil (or dimethyluracil) are irradiated under nitrogen with kilorad doses of X-rays, the 6-hydroxy, 5-hydro derivatives (water addition products) are formed (Figure 10-6). These same compounds are formed when uracil (or dimethyluracil) are irradiated in solution with UV (2537 Å). After the irradiation of frozen solutions of thymine with megarad doses of cobalt-60 γ-rays, a product has been isolated which is apparently identical to the thymine dimer originally produced by UV. Thus, under certain conditions X-rays and UV can produce the same radiation products. The extreme doses of γ-rays used in the last example, however, make one wonder about the relevance of this observation to radiation biology. It has also been observed that photochemically produced thymine dimers can be split by ionizing radiation to yield thymine and other products.

## 10-8. RADIATION CHEMISTRY OF DNA

The most prominent effects produced in DNA are hydrogen bond breakage, degradation of the bases and chain scission. It is the latter lesion which best correlates with biological inactivation. Even

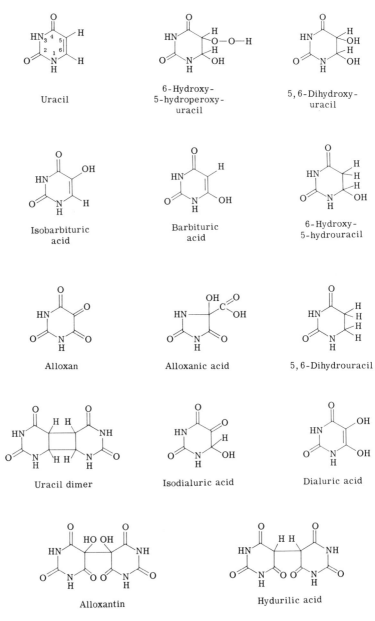

FIGURE 10-6. The structure of various derivatives of uracil. [From K. C. Smith and J. E. Hays, *Radiation Res.* **33**, 129 (1968).]

though DNA chain breaks are produced by X-irradiation, very little inorganic phosphate is liberated. This stability of the phosphate groups to radiation has been the basis for an assay for chain breaks. A chain break at a phosphate bond should yield a molecule of monoesterified phosphate. The amount of inorganic phosphate liberated from irradiated DNA by an enzyme specific for mono-esterified phosphate should give a measure of the number of chain breaks occurring adjacent to phosphate groups. This method would not measure chain breaks produced as a consequence of the disruption of deoxyribose residues and there is evidence of extensive sugar damage in irradiated DNA.

Furthermore, the evidence that the polynucleotide rejoining enzyme will not repair chain breaks produced in DNA irradiated *in vitro* suggests that the chain breaks are more complicated than a clean scission at a phosphate bond. This enzyme will rejoin single chain breaks containing juxtaposed 5'-phosphate and 3'-hydroxyl groups (as can be produced with pancreatic deoxyribonuclease). The detection of extensive sugar damage and the observation that the majority of the monoesterified phosphate produced by X-irradiation of DNA is not 5'-phosphate, helps to explain the lack of repair of X-ray induced chain breaks by the polynucleotide rejoining enzyme (ligase) alone.

Summers and Szybalski developed an assay for single strand breaks that makes use of DNA molecules that contain one cross-link between the two strands, as produced by treatment with the antibiotic mitomycin. Such crosslinked DNA will heat denature at elevated temperatures but will again form double-stranded DNA on cooling. This renatured DNA will again band in a CsCl density gradient at the density position for double-stranded DNA. If a single strand break is introduced into this molecule by X-irradiation, when the sample is heated and cooled a single-stranded piece of DNA will become disassociated from the main double-stranded piece and the two pieces will band at different positions in the CsCl gradient. Single-strand breaks have been observed by this method in DNA exposed to doses of X-rays as low as 250 R, with natural variation in the base composition of the DNA or 5BU substitution having little influence, and cysteine exerting a protective effect.

More recently the alkaline sucrose density gradient technique developed by McGrath and Williams has been used to determine the presence of single strand breaks in irradiated DNA. This technique combines two known facts: alkali separates the two strands of DNA and the rate of sedimentation in a sucrose gradient is proportional to molecular weight. Single chain breaks produced in DNA by X-irradiation would be detected as a decrease in sedimentation rate compared to the unirradiated sample.

With this technique McGrath and Williams first showed that during incubation in growth media, a radiation resistant strain of *E. coli* could repair radiation-induced single strand breaks (i.e., the molecular weight of the irradiated DNA returned essentially to the normal value) while a radiation sensitive strain could not rejoin these breaks (Figure 10-7).

Kaplan has carried this technique further and shown that the survival of a resistant strain is not correlated with the production of single strand breaks (which is logical since they are efficiently repaired) but, rather, lethality is inversely correlated with the production of double chain breaks (assayed in neutral sucrose gradients). Double chain breaks do not appear to be repaired in *E. coli* strains (Figure 10-8). This correlation between lethality and the production of double chain breaks persisted when the cells were sensitized by the incorporation of 5-bromouracil in place of thymine (Figure 10-9) or when the cells were protected by irradiation in the presence of cysteamine.

10-9. THE INTERACTION OF X-RAYS AND UV

There is a synergistic interaction of X-rays and UV on the survival of certain strains of bacteria. To demonstrate this, bacteria are killed to a certain survival level with one type of radiation and then a survival curve is determined for the second type of radiation (Figure 10-10). Preirradiation with X-rays removes the shoulder of the UV survival curve for *E. coli* B/r but does not change the final slope. It is difficult to tell from these data whether there has been synergism or just additivity. In the reciprocal experiment where the survival curve is exponential, it is clear that prior treatment with UV renders the surviving fraction more sensitive to subsequent X-ray treatment. There is an upper limit to this sensitization by prior UV as shown in curves 5 and 6. These curves are parallel to one another and are about 3 times as steep as curve 1.

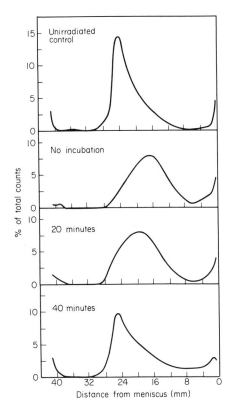

FIGURE 10-7.   Sedimentation in alkaline sucrose gradients of tritium labeled DNA from control and irradiated (20 kR) *E. coli* B/r. The lower panels indicate the shift in sedimentation observed when irradiated cells are incubated in growth medium for 20 to 40 minutes after irradiation, respectively, before being placed on the gradient. [Adapted from R. A. McGrath and R. W. Williams, *Nature* **212**, 534 (1966).]

Synergism has also been shown in *E. coli* B/r between nitrogen mustard and X-rays and nitrogen mustard and UV.

It has been suggested that this enhanced killing (synergism of X-rays and UV) is due to the failure of recovery processes because the damaged sites are either too numerous or too closely spaced in the DNA to allow repair (or the interaction produces damage that cannot be repaired). The main evidence to support this concept was the observation that this synergistic response is not shown by

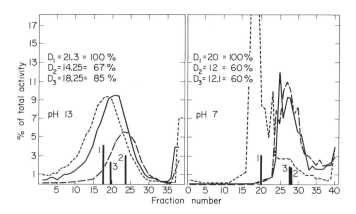

FIGURE 10-8.   Sedimentation patterns of DNA from irradiated and control cells
(*E. coli* K-12; JE850). In pH 13 sucrose gradients, there is a sharp decrease in sedimentation rate of alkali-denatured DNA after 20 krads, with restoration almost to
normal during 40 minutes of reincubation at 37°C. A marked change in sedimentation behavior at pH 7 is also seen with undenatured DNA after 20 krads, but there
is no evidence of repair during reincubation. ----, control + 0 krad; − −, control
+ 20 krad; ——, control + 20 krad + 40 minute incubation. [From H. S. Kaplan,
*Proc. Natl. Acad. Sci. U.S.* **55**, 1442 (1966).]

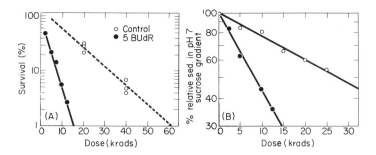

FIGURE 10-9.   (A) Dose survival curves for control and BUdR-grown cells reveal
a three-fold increase in slope after BUdR incorporation into DNA. (B) A semilog
plot of relative sedimentation distance of the DNA from these cells (*E. coli* K-12)
vs. X-ray dose yields exponential curves with a similar three-fold increase in slope
for BUdR-DNA vs. that from control cells. [From H. S. Kaplan, *Proc. Natl. Acad.
Sci. U.S.* **55**, 1442 (1966).]

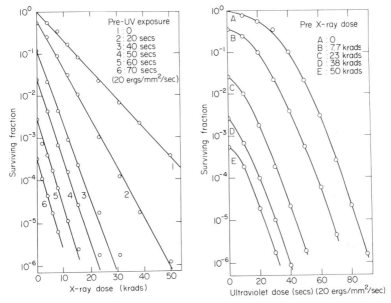

FIGURE 10-10. Synergistic interaction between UV and X-rays. [From R. H. Haynes, *in* "Physical Processes in Radiation Biology" (L. Augenstein, R. Mason, and B. Rosenburg, eds.), p. 51. Academic Press, New York, 1964.]

cells (e.g., *E. coli* $B_s$) that are known to be deficient in dark repair mechanisms for UV damage. However, Smith and Ganesan have shown that the two radiations interact to yield protection for *E. coli* $B_s$. At the maximum practical dose of UV used (17.5 ergs/mm², leaving a surviving fraction of $10^{-3}$), the slope of the subsequent X-ray survival curve was less by a factor of 0.66 than for cells without a prior UV treatment.

Although several hypotheses have been advanced (and some tested) to explain why X-rays and UV interact synergistically, with additivity, or to yield protection in different strains of *E. coli*, there is no clear understanding of the mechanisms involved. It is unlikely that studies on the interaction of X-rays and UV on survival will yield meaningful conclusions until the response of cells to the separate radiations is better understood.

## GENERAL REFERENCES

Z. M. Bacq and P. Alexander, "Fundamentals of Radiobiology," 2nd ed. Pergamon Press, Oxford, 1961.

M. Errera and A. Forssberg, eds., "Mechanisms in Radiobiology," Vol. 1. Academic Press, New York, 1961.

Fundamental aspects of radiosensitivity. *Brookhaven Symp. Biol.* **14**, (1961).

J. J. Weiss, Chemical effects of ionizing radiations on nucleic acid and related compounds. *Prog. Nucleic Acid Res. Mol. Biol.* **3**, 103 (1964).

J. W. T. Spinks and R. J. Woods, "An Introduction to Radiation Chemistry." Wiley, New York, 1964.

L. G. Augenstein, R. Mason, and B. Rosenburg, eds., "Physical Processes in Radiation Biology." Academic Press, New York, 1964.

L. G. Augenstein, R. Mason, and M. Zelle (eds.), *Advan. Radiation Biol.* **1**, (1964).

M. Ebert and A. Howard (eds.), *Current Topics Radiation Res.* **1**, (1965).

G. R. A. Johnson and G. Scholes (eds.), "The Chemistry of Ionization and Excitation." Taylor & Francis, London, 1967.

A. P. Casarett, "Radiation Biology." Prentice-Hall, Englewood Cliffs, New Jersey, 1968.

# SUBJECT INDEX

Note: The letter G among the page citations refers to the Glossary found at the front of the book (page xi).

217

# Y

# Z